The Constitution of Science

How can science be protected, by whom and at what level? If science is valued positively as the incubator of the most successful solutions to representational problems of reality as well as the basis of the most effective interventions in the natural and social world, then its constitutional foundations must be protected. This book develops a specific normative outlook on science by introducing the idea of a 'Constitution of Science'. Scientific activities are special kinds of epistemic problem-solving activities unfolding in an institutional context. Those institutions of science which are of the highest generality make up the 'Constitution of Science' and are of fundamental importance for channelling the scientific process effectively.

C. MANTZAVINOS is Professor of Philosophy at the University of Athens and an elected member of the Academia Europaea and the International Academy for the Philosophy of the Sciences. He has taught at Freiburg, Stanford and Witten/Herdecke, was a senior fellow at the Max Planck Institute for Research on Collective Goods, Bonn, and held visiting appointments at Harvard, Princeton, and the Maison des Sciences de l'Homme, Paris. He is the author of *Wettbewersbstheorie* (Duncker & Humblot, 1994), *Individuals, Institutions and Markets* (Cambridge University Press, 2001), *Naturalistic Hermeneutics* (Cambridge University Press, 2005), *Explanatory Pluralism* (Cambridge University Press, 2016), *A Dialogue on Explanation* (Springer Nature, 2018), *A Dialogue on Institutions* (Springer Nature, 2021) and the editor of *Philosophy of the Social Sciences* (Cambridge University Press, 2009).

'In a time when disinformative challenges to the results of science are widespread, Chrys Mantzavinos offers this careful, wide-ranging, ambitious, multi-pronged investigation into the many necessary and perhaps sufficient conditions to foster healthy scientific inquiry. Normativity emerges as a key to the solution in a context where problem-solving is central, social interactions are essential activities, and diverse values must be respected. Free expression, critical discussion, open access are principles that allow discovery, justification and applications to take place in this innovative philosophical account.'

James Alt, Frank G. Thomson Professor of Government Emeritus, Harvard University

'A work of high intellectual ambition that offers an account of what makes science science from philosophical, social and economic perspectives. Mantzavinos never forgets that science and the societies in which it has thrived have been diverse and constantly evolving. Yet he offers an account of science – its constitution – that aims to capture its core values, both moral and epistemic.'

Lorraine Daston, Director Emerita, Max Planck Institute for the History of Science

'Chrysostomos Mantzavinos has offered us a bold new perspective on the scientific enterprise. He views science as a polity with an unwritten constitution, and shows how this idea can cast new light on familiar philosophical questions. After decades during which the social character of science has been proposed and debated, philosophy informed by social theory finally steps in to take the notion seriously. This book should inspire a new genre of fertile discussions.'

Philip Kitcher, John Dewey Professor of Philosophy Emeritus, Columbia University

'This outstanding book offers an innovative view of science as an unplanned, evolving practice with its own flexible norms; shows how well it has served us in enforcing a varied batch of constraints on the search for knowledge; and identifies the principles under which our states should allow it to honour its informal constitution.'

Philip Pettit, L. S. Rockefeller University Professor of Human Values, Princeton University, and Distinguished Professor of Philosophy, Australian National University

The Constitution of Science

C. MANTZAVINOS
University of Athens

Shaftesbury Road, Cambridge CB2 8EA, United Kingdom

One Liberty Plaza, 20th Floor, New York, NY 10006, USA

477 Williamstown Road, Port Melbourne, VIC 3207, Australia

314–321, 3rd Floor, Plot 3, Splendor Forum, Jasola District Centre, New Delhi – 110025, India

103 Penang Road, #05–06/07, Visioncrest Commercial, Singapore 238467

Cambridge University Press is part of Cambridge University Press & Assessment, a department of the University of Cambridge.

We share the University's mission to contribute to society through the pursuit of education, learning and research at the highest international levels of excellence.

www.cambridge.org
Information on this title: www.cambridge.org/9781009509183

DOI: 10.1017/9781009509190

When citing this work, please include a reference to the DOI 10.1017/9781009509190

First published 2024

A catalogue record for this publication is available from the British Library

Library of Congress Cataloging-in-Publication Data
Names: Mantzavinos, Chrysostomos, author.
Title: The constitution of science / C. Mantzavinos, University of Athens, Greece.
Description: Cambridge, United Kingdom ; New York, NY, USA : Cambridge University Press, 2024. | Includes bibliographical references and index.
Identifiers: LCCN 2024001808 | ISBN 9781009509183 (hardcover) | ISBN 9781009509176 (paperback)
Subjects: LCSH: Science – Philosophy. | Science – Methodology.
Classification: LCC Q175 .M3465 2024 | DDC 501–dc23/eng/20240509
LC record available at https://lccn.loc.gov/2024001808

ISBN 978-1-009-50918-3 Hardback
ISBN 978-1-009-50917-6 Paperback

To

PETROS GEMTOS

ἀλλά μοι ψεῦδός τε συγχωρῆσαι καὶ ἀληθὲς ἀφανίσαι οὐδαμῶς θέμις.

Πλάτων, Θεαίτητος, 151d

τὸ μὲν γὰρ λέγειν τὸ ὂν μὴ εἶναι ἢ τὸ μὴ ὂν εἶναι ψεῦδος, τὸ δὲ τὸ ὂν εἶναι καὶ τὸ μὴ ὂν μὴ εἶναι ἀληθές.

Ἀριστοτέλης, Μετὰ τὰ φυσικά, 1011b25

Contents

Acknowledgements

It is my pleasant duty to thank all the colleagues and friends who have supported my work on this book. I am grateful to Petros Gemtos who has originally sparked my admiration for science and has taught me the significance of institutions for social life. The work and discussions with Hans Albert, who has recently passed away, has been a constant inspiration for me over many years. I want also to express my gratitude to Herbert Keuth, my teacher in Tübingen, for all that I have learned from him.

Philip Kitcher has provided detailed and extremely valuable criticisms on the first and second drafts of the manuscript; beyond my obvious debt to his own philosophical work, I am grateful for all his help and support over the last two decades. I have profited enormously from discussions about the main ideas of the book with Philip Pettit at Princeton; his philosophical work and personality have been the source of constant inspiration for me. Martin Carrier has read and commented on early material when the book was at its infancy and pressed me to clarify my arguments; I am very grateful for our memorable discussions in Bielefeld and in Athens, and I am honoured by his friendship.

I am indebted to Alexander Nehamas for his support of the project; I would also like to thank him for our wonderful exchanges on a great range of philosophical topics over many years.

I am grateful to James Alt for his support of the project and for my interaction on the role of institutions in social life.

I want to thank my dear colleagues and friends in Athens. Theo Arabatzis has provided detailed comments that have improved considerably the manuscript – I am very thankful for all our exchanges over many years. Panagiotis Thanassas has saved me from a series of errors in the original draft of the manuscript for which I am very thankful. I have learnt a lot about metaethics from Stelios Virvidakis, who has also provided precious comments on the original version of the chapter on normativity for which I am grateful. I want to thank Anthony Hatzimoysis who has also commented extensively on the original draft of the chapter on normativity and made a series of extremely important suggestions that helped me improve the quality of the chapter. I would also like to thank particularly my colleagues Stathis Arapostathis Jean Christianidis, Costas Dimitracopoulos, Stavros Drakopoulos, George Gotsis, Aris Hatzis, Katerina Ierodiakonou, Stavros Ioannidis, Pavlos Kalligas, Doukas Kapantais, Vassilios Karakostas, Vasso Kindi, Eleni Manolakaki, Stathis Psillos, Yannis Stephanou and Aristotelis Tympas for many helpful discussions.

I had the luck to serve as Mercator Fellow at the DFG Research Training Group "Ethics and Epistemology of Science" at Bielefeld University/Leibniz University Hannover for a number of years, and I am grateful to all the faculty members, postdocs and doctoral students of this wonderful group for their valuable feedback. I would like to especially thank Torsten Wilholt for being a wonderful host and for all that I have learnt from him during our exchanges. I would also like to particularly thank the Principal Investigators of the group: Alkistis Elliott-Graves, Uljana Feest, Mathias Frish, Dietmar Hübner, Marie Kaiser, Thomas Reydon and Ralf Stoecker.

I thank THE NEW INSTITUTE, Hamburg, for a fellowship in the Spring Term 2023, which allowed me to make many substantial improvements on the manuscript. I would like to especially thank Ruth Chang and Kit Fine for discussions on the ideas contained in the book and Kit especially for his valuable comments on Chapter 4. I want also to thank Avram Alpert, Madhulika Banerjee, Anthea Behm, Carole Bloch, George Ellis, Jim Guszcsa, Geoffrey Harpham, Max Krahé, Bruno Leipold, Tobias Müller and Maki Sato for useful exchanges of ideas. My special thanks go to the librarians, Christiane Müller and Britta Neumann, for their wonderful support. While in Hamburg I have profited a lot from interactions with Stefan Voigt – I am very thankful to him for all his comments and his goodwill. I have had very fruitful discussions on various philosophical topics with

Judith Simon for which I thank her. Many thanks also to Thomas Krödel for helpful discussions and for his valuable comments on the chapter on normativity.

Many sincere thanks to Emmanuel Picavet for being my wonderful host at the Maison des Sciences de l'Homme, Paris, in 2018 and for our exchanges on institutional analysis for many years. Thanks also to Daniel Andler, Alban Bouvier, Gilles Campagnolo, Pierre Demeulenaere, Jean-François Kervegan, Gianluca Manzo and Christian Walter for valuable discussions.

I have benefited from discussions with Bernhard Nickel, Ned Hall, Alex Byrne, Sally Haslanger and Panagiotis Roilos during my short visit at Harvard in 2023 for which I want to thank them.

I would like to thank Thomas Schmidt for our memorable discussions in Berlin and Athens and his detailed comments on the chapter on normativity that helped me avoid a series of errors. I would also like to thank Stefan Magen for his invitation to Bochum and the valuable discussions that we have had over a broad array of philosophical issues since our time at the Max Planck Institute for Research into Collective Goods in Bonn. For insights offered on different occasions, I would like to particularly thank Pablo Abitbol, Pantelis Analytis, Justin Biddle, Alain Bresson, Tim Crane, Lorraine Daston, Immaculada De Melo-Martin, Pascal Engel, James Evans, Dan Garber, Gerd Gigerenzer, Francesco Guala, Jacob Habinek, Jens Harbecke, Catherine Herfeld, Geoffrey Hodgson, Paul Hoyningen-Huene, Ian Jarvie, Stathis Kalyvas, Uskali Mäki, Johannes Mierau, Carl-David Mildenberger, Eleonora Montuschi, David Papineau, Alex Paseau, Samuli Reijula, Kristina Rolin, Jacob Stegenga, Michael Strevens and Julie Zahle.

I am very grateful to Panagiotis Karadimas and Jonas Lipski for many very helpful comments on the first draft of the manuscript. Leonidas Tsilipakos has also provided valuable comments on the entire manuscript for which I want to thank him. I am also grateful to the participants of my graduate seminar in Athens for their comments, especially Vasilis Daniil, George Mouzakis, Nikolaos Sorolis, Foivos Syrigos and Ioanna Tsitsou.

I would very much like to thank two anonymous reviewers for a series of extremely useful comments.

I am indebted to my editor at Cambridge University Press, John Haslam, for his goodwill and wonderful support of the project. I am also indebted to Hilary Gaskin for her significant backing of the project. I owe them both a very special thanks. I would like to thank Carrie Parkinson for her guidance through the final stages until the publication of the book. I thank also Christian Green and Swati Kumari for all their help in the production phase of the book.

Darrell Arnold, as always, has provided many important comments and has improved the language in many chapters of the book. I am very grateful to him for his help also this time.

I would like to thank SAGE for permission to use material from the essay "Institutions and Scientific Progress" which was published in *Philosophy of the Social Sciences*, vol. 51, 2021, pp. 243–265 and WILEY for permission to use material from the essay "Science, Institutions and Values" which was published in the *European Journal of Philosophy*, vol. 29, 2021, pp. 379–392. Figure 1 on The Problem Solving Framework has been reproduced from my book *Individuals, Institutions, and Markets*, Cambridge University Press, 2001, p. 41.

My greatest debt is to my family, especially to my wife Georgia for her love and affection during all those years that we have been together.

Introduction

Science as an arena of epistemic activities has grown organically during a long process of cultural evolution. It is a major cultural achievement that was planned by no individual mind but emerged spontaneously as the unintended outcome of interaction between individuals engaging in epistemic activities. Science is a human endeavour, permanently unfinished, a project of humanity of astonishing range and success. Such a complex phenomenon excites our intellectual curiosity inducing us to adopt a *descriptive stance* explaining its emergence and its mode of operation and incites our admiration inducing us to adopt a *normative stance* providing good reasons for supporting its further existence and for securing its accomplishments.

In this book I will develop a specific outlook on science from within the normative stance by virtue of introducing the idea of a Constitution of Science. Scientific activities are special kinds of epistemic problem-solving activities unfolding in an institutional context. The scientific enterprise is a social process unfolding within an intricate institutional framework that structures the daily activities of scientists and shapes their outcomes. Those institutions of science which are of the highest generality make up the Constitution of Science and are of fundamental importance for channelling the scientific process effectively. If science as the incubator of the most successful solutions to problems of representation of reality and as the basis of the most effective interventions in the natural and social world is valued positively, then its constitutional foundations must be protected. How exactly, by whom and at what level are the questions that I would like to address in this book.

1 The Scaffolds Humans Erect on Science

Imagine that we are given the possibility of constructing a building on an island without a master plan and without a specific time horizon for its completion. Anyone of us may bring as many bricks as (s)he can to construct the building. Some of us would start putting one or more bricks in this collective enterprise, and there would be discussions, agreements and disagreements on how to fit the bricks together so that the construction can stand on its own. For the building to take shape scaffolds will be needed. And as the building gets taller and taller, the old scaffolds will have to get adjusted or entirely new scaffolds will have to be created using the materials of the old ones. You cannot climb higher and add more building blocks – bricks, that is – unless you mount the steps of the scaffolds already erected. And you cannot make the building more useful or more beautiful by painting it, for example, unless you use the scaffolds. Earthquakes might damage the building or even destroy it, but once you make use of the scaffolds you can rebuild it.

Why are we constructing the building? Some of us because we want to stand higher and desire to enjoy a better view. Others because they think that those on the ground will admire them for their achievement to stand higher. Others because they expect that by telling different people on the ground what they can see when on the building, they will give them money for what they will hear. Still others because of the mere challenge of participating in the collective enterprise.

As the process of construction evolves, those participating get more and more accustomed to using the scaffolds, to adjusting them when necessary and to extending them to construct side buildings next to the main building. The buildings provide possibilities for enjoying further views of reality (of the surroundings, the sea, the sky, the island on which the edifice is erected, but also the other islands), depending on where we stand. And when we move ourselves on the same building or move to another building of the edifice, we get a thoroughly new perspective of the surroundings.

But is there any kind of synthesis? Is this plurality of views all that there is? The very fact that buildings are being constructed and scaffolds are being erected and that there are people on the buildings is what makes up the synthesis; it is what we call 'science'. And this is fundamentally different than what is going on the ground. All of them on the buildings have a better view than the ones on the ground; and if they can climb on different scaffolds and reach different buildings, they can enjoy diverse views. The complex of men, scaffolds and buildings *is* the synthesis.

Do we want to retain this synthesis? How exactly? Who can protect it and at what level? There will always be people on the ground who will challenge the exclusivity of the view from the buildings. If you are on the ground, you can also enjoy a nice view, they will say, equally nice with the view from the buildings, so why should one take the trouble of erecting scaffolds and constructing buildings? The best answer to them will be to give them a hand and help them jump on one of the buildings. Some will do so and change their mind once there. Others will decline and will continue protesting. As long as most people on the ground will tolerate the process of construction and find, for whatever reason, the stories told by those on the buildings valuable, the constructors will be left alone to extend and cultivate the edifice further – they might even be supported in their endeavours. The edifice of science will continue to exist and be developed further, only if the scaffolds are appropriately adapted over time. The main precondition for this to happen is that the values of those on the construction side and those on the ground are harmonised in a congruous way. Science can flourish only if suitably harmonious values uphold it.

2 Science and Values

Is there one value which is more important than the others? Truth perhaps? This was the suggestion of Max Weber, the original advocate of a 'value-free science', as well as the many who have followed him.[1] What we value most in statements and theories delivered by science is that they offer us a true description of the natural and social world or at least a description that is closer to truth than all other alternatives. The scientific statements themselves, even the ones that refer to the social world, *can* be formulated in such a way that they are not of a normative character. This view, originally proposed during the *Werturteilsstreit*, was subsequently defended by Karl Popper and Hans Albert with new arguments during the legendary *Positivismusstreit* in Germany, rather wittily by Henri Poincaré in France, and from logical-positivistic premises by Hans Reichenbach in the Anglo-Saxon world.

Isaac Levi (1960) has broadened the discussion here by suggesting that canons of inference limit the values scientists can and should use in their evaluation of scientific hypotheses. He has defended the view that, even if truth is not the sole operative value of science, *epistemic values* still make for sufficient criteria with respect to scientific thinking. Epistemic values are the values that promote the attainment of truth[2] and can be distinguished from non-epistemic values (McMullin, 1983/2012; Lacey, 2005). The "underdetermination thesis," that is the thesis that evidence cannot prove the truth of a theory or, in other words, that a theory, hypothesis or model is *in principle* underdetermined irrespective of the amount of evidence that we might be able to collect, questioned the exclusive focus on epistemic values.[3] The debate on the argument from inductive risk according to which the decision whether to accept or reject a hypothesis must often, if not always, take into consideration not only the likelihood of error, but also how bad are the consequences of error when the hypothesis is accepted or rejected has also challenged the priority of the epistemic values.[4]

Are truth and other epistemic values such as accuracy, consistency, broad scope, simplicity and fruitfulness the most important values?[5] This is a position that can be defended as long as one focusses on the *outcomes* of the scientific endeavour – mainly theories and models. If one also takes into account the social *process*[6] that yields these outcomes, then a series of other values become not only important, but indeed constitutive of science: if one is not free to engage in scientific activity, for example, then there are no outcomes in the first place to be appraised invoking epistemic values. It seems that non-epistemic values such as freedom, honesty and integrity[7] are also constitutive of the collective scientific endeavour.

Do we possess an external criterion guaranteeing the priority of a single value or a set of values vis-à-vis the rest? Consider the principle of Roman law 'fiat iustitia et pereat mundus' – 'let justice be done though the world shall perish'. Should justice be given *absolute* priority? The Philalethes in Schopenhauer's dialogue *Über Religion* endorses the principle 'vigeat veritas et pereat mundus' – 'let truth prosper, though the world shall perish' (Schopenhauer, 1851). Should truth be given *absolute* priority? There are justifications for relying more or less on a single value or a set of values, but is there also an *ultimate* justification that may compel us as agents endowed with reason to settle on the priority of one or another set of values? Humans used to believe in the existence of external criteria and the possibility of an ultimate justification of a certain set of values for centuries, indeed millennia. Such a conviction was held in parallel with a quest for certainty in epistemology and ethics. The *principle of sufficient reason* nicely encapsulated this belief. However, the discussion of Agrippa's five tropes in the context of ancient scepticism had already considerably shaken the faith in the possibility of attaining certainty.[8]

In the modern discussion,[9] it was the German philosopher Hans Albert (1968/1985) who forcefully argued against the vain quest for certainty in epistemology and in ethics stressing that the quest for an ultimate justification for our convictions and principles leads to a situation with three alternatives which are all unacceptable: the Münchhausen Trilemma[10] (p. 18): "[I]f one demands a justification for *everything*, one must also demand a justification for the knowledge to which one has referred back the views initially requiring foundation. This leads to a situation with three alternatives, all of which appear unacceptable: in other words, to a trilemma which, in view of the analogy existing between our problem and one which that celebrated and mendacious baron once had to solve, I should like to call the *Münchhausen trilemma*. For, obviously, one must choose between

1. an *infinite regress*, which seems to arise from the necessity to go further and further back in the search for foundations, and which, since it is in practice impossible, affords no secure basis;
2. a *logical circle* in the deduction, which arises because, in the process of justification, statements are used which were characterized before as in need of foundation, so that they can provide no secure basis; and, finally,
3. the *breaking-off of the process* at a particular point, which, admittedly, can always be done in principle, but involves an arbitrary suspension of the principle of sufficient justification."

Since the first two options are normally considered unacceptable, one usually suspends the argumentation at some point using terms such as "intuition" and "self-evident." The ultimate "given" is then baptised as the Archimedean point by the respective philosopher supposedly providing the secure foundation of the justificatory procedure. "Justification by *recourse to a dogma*" (Albert, 1968/1985, p. 19) is then the outcome.[11]

If there is no Archimedean point then, what is the alternative? An alternative is to accept *fallibilism*, embrace *pluralism* and adopt the *principle of critical examination* instead of the principle of sufficient justification. This alternative stance mandates that value judgments can all be accepted as hypotheses amenable to criticism. Value pluralism is what one has to deal with and a trade-off between values becomes mandatory: we can more easily find the truth about the maximum amount of mustard and nerve gas that a human body can endure by conducting experiments, but is the pain inflicted upon human subjects in such experiments acceptable? Who should make the trade-offs and at what level? Values encapsulate normative resources at the highest level, and they are important as guides to interact with others in a social group. A monad inhabiting a world for himself or herself would not need any values. But who should undertake the trade-offs between values in a society which is also organised as a political community? This is a problem that *can be* and *is de facto* addressed in the form of a constitutional issue: what are the highest institutional principles in an organised polity that should regulate the functioning of science, so that all the values that we deem important are appropriately reflected and sufficiently traded-off?[12]

In order to adequately answer this question, a lot of preparatory work is necessary. First of all, it should be made clear that our talk about trade-offs between values is a liberal one, a useful, though dramatic simplification that should not mislead us to adopt any kind of reification of values. Values should not be conceptualised as things, as detached entities which are mind independent. Values encapsulate normative resources at the highest and most abstract level, and they are eminently important as guides to interact with others in a social group. Since so much of my argument hinges upon the role of values in science, a thorough treatment of the issue of normativity becomes necessary. Normativity is an issue of fundamental importance not only for practical philosophy (e.g., for understanding rules and institutions) but also for theoretical philosophy (e.g., for understanding epistemic norms or the appraisal of scientific theories). The aim of Chapter 3 is to develop a workable naturalistic account of normativity informed by the sciences.

3 Normativity

Long before they become able to engage themselves in speculations and to create a sophisticated *Weltanschauung*, human beings have to encounter practical problems successfully in order to secure their further existence. Their principal concern is to *act* in order to master diverse problems in their natural and social environment. Human action requires forming mental models for the solution of practical problems that provide an orientation for dealing with inanimate and animate nature. One may presume that a great part of human cultural evolution was characterised by a natural value-Platonism of the everyday view of the world, which was facilitated by a linguistic fusion of facts and values.[1] A sociomorphic interpretation of the world occurs in many civilisations and manifests itself in diverse views: for example, in the view that the world is a human body or in the view that the origin of the universe is analogous to human conception, birth and fertility.[2] The distinction between facts and values is not even considered in these first stages of cultural evolution, since anthropomorphism is transferred quasi-automatically to the cosmos as a whole.

One could plausibly assume that the thorough analysis of this development by the modern social sciences would have led to a general consensus also in moral philosophy and that the disenchantment of the world – to use the classic phrase of Max Weber[3] – would have become a universally accepted thesis. However, this is not the case. In the philosophical discussion of the last decades, the position has gained a foothold according to which there is a more or less well-identifiable, partly detached domain of values, which is not necessarily hypostatised, but which supposedly belongs to the furniture of the world. This discussion is commonly conducted using the vocabulary of "moral realism," and it has in the meantime generated subtly nuanced formulations and argumentations.[4]

I will not address contemporary metaethical positions, which after an initial phase during which they focused on arguments for or against moral realism,[5] have subsequently centred on the nature of normativity.[6] I will only state that the subtlety of positions and the ingenuity of argumentations are impressive – expressed in a philosophical style that ceased to be baroque, intended for outdoor use, and has taken up features of rococo, which is at home mainly indoors. And I will proceed with developing a workable naturalistic account of normativity informed by the deliverances of the sciences.

3.1 "THE DOMAIN OF THE NORMATIVE" AS VIEWED FROM THE SCIENCES

It is commonplace in the disciplines that focus on *biological evolution* that nonhuman animals can form mental representations, feel emotions, and that exchange processes and cooperation take place in animal groups.[7] There are, of course, intense debates about a series of issues, but there is a consensus that mechanisms of evaluation and emotions proper are available in animals[8] – a minimalist thesis, unaffected by the answer to the question about the extent of animal abilities to recognise wishes and thoughts of their conspecifics.[9] Theories of *cultural evolution* show how rules of conduct emerge in processes of mutual adaptation and how they evolve over time.[10] The evolution of normative rules enabling collaboration in the human species goes hand in hand with the evolution of cognitive skills of joint intentionality permitting the successful overcoming of ecological challenges.

Neurological studies show that certain aspects of the process of emotion and feeling are indispensable for human decision-making,[11] and that the inevitable fact about the phenomena of emotion, feeling and indeed consciousness is their body relatedness.[12] There is, from a neurological point of view, nothing mysterious about emotions, in the sense that they can be dealt with as phenomena taking place within the brain, which can be researched by a series of techniques developed by neuroscience. And a series of studies conducted by *cognitive psychologists* building on the insight that humans avail of evolved brains show that gut feelings – our intuition – very often guide human decision-making and do so very successfully. What we call "intuition" is, again, nothing mysterious, but can be analysed as a series of *Simple Heuristics That Make Us Smart* (Gigerenzer),[13] that is rules of thumb that are the outcome of evolved capacities and lead to decisions in the absence of long deliberation.[14] The environment, of course, plays a crucial role in helping trigger these heuristics, but the mind as the collection of cognitive abilities, as an adaptive toolbox,[15] reflects both the result of evolution and the personal experiences of an individual within her specific environment; and it can give rise to both long and intricate deliberations *and* intuitive solutions to problems, the success depending on the match between the respective cognitive ability and the structure of the problem encountered in the natural or social environment.

The subjective theory of value has been the cornerstone of *modern economic theory* since the marginal revolution in the 1870s.[16] It rests on the fundamental insight, already championed by Carl Menger,[17] that economic subjects proceed in constant evaluations of goods and services in markets, and it is considered simple textbook economic knowledge that the source of these evaluations is the individual mind. Notwithstanding criticism from all possible alternative approaches to the neoclassical orthodoxy, such as behavioural economics[18] and institutional economics,[19] there is a consensus about this point.

Finally, *social theory* analysing the emergence, functioning and causal force of social norms from different viewpoints also never includes any mechanisms of discovery of them in the explanations it provides. On the contrary, processes of bargaining and collective action are normally problematised, as is prominently done in Jon Elster's *The Cement of Society*.[20] Alternatively, mechanisms of sanctioning the deviators are highlighted[21] *or* processes of social learning are emphasised during which social norms are diffused in the group,[22] *or* the stability of norms is analysed drawing on computer simulations,[23] etc. None of these approaches in social theory postulates the existence of social norms outside the brains of the interacting individuals.

Taking stock of the scientific research in the "domain of the normative" as presented in this extremely condensed review, that is, of the research on a series of diverse phenomena of normativity on which very diverse scientific disciplines focus their attention, one can record that none of these phenomena are located beyond mental activity, whether individual or shared. Science teaches us that emotions and feelings are neurologically embedded in brains, are manifested by the functioning of diverse mechanisms and thus exist as constitutive parts of brains. Finally, science teaches us that evaluations as revealed in decision-making procedures and rule-following are the outcome of a complex interaction during which more general brain networks are activated, so that both emotions and purely cognitive processes give rise to their manifestation. I will attempt a synthesis of some of these results of the sciences by means of presenting a problem-solving framework aiming at shedding light on the most important phenomena of normativity: choice, individual rule-following, emergence of social normative rules, and the interaction between the individual and its environment.

3.2 PROBLEM-SOLVING, CHOICE AND RULE-FOLLOWING

The central claim of the problem-solving framework is that all human activity can be viewed as problem-solving activity. Human beings are constantly confronted with problems, and they mobilise their cognitive and emotional resources in order to solve them. Newell and Simon provide a useful first approximation of the view to be developed further here: "A person is confronted with a problem when he wants something and does not know immediately what series of actions he can perform to get it. The desired object may be very tangible (an apple to eat) or abstract (an elegant proof of a theorem).

It may be specific (that particular apple over there) or quite general (something to appease hunger). It may be a physical object (an apple) or a set of symbols (the proof of a theorem). The actions involved in obtaining desired objects include physical actions (walking, reaching, writing), perceptual activities (looking, listening), and purely mental activities (judging the similarity of two symbols, remembering a scene, and so on)."[24]

Problem-solving can be conceptualised as a process of searching through a state space. One can define a problem as an initial state, by one or more goal states to be reached by a set of operators that can transform one state into another and by constraints that a solution must meet.[25] The problem-solving procedure can be conceptualised, thus, as a series of steps in which operators are applied so that the goal states can be reached. Problems can in principle be of two kinds: theoretical and practical. Declarative knowledge – "knowledge that" – emerges as the outcome of solving "theoretical problems," procedural knowledge – "knowledge how" – emerges as the outcome of solving "practical problems." This categorical distinction, first introduced by Ryle,[26] is now broadly accepted in diverse scientific disciplines and can be crystallised with the help of a simple criterion: the specific communicability of knowledge. "Knowledge that" can be communicated by means of language using a series of symbols, whereas skills and arts cannot be communicated by symbolic languages, but through learning by example, that is, learning by imitation.[27] The distinction between declarative knowledge and procedural knowledge is a largely accepted one in cognitive science, and it is generally acknowledged that learning facts is very different than acquiring skills, even if a problem-solving mechanism underlies both of them.[28]

The role of emotions in the problem-solving process can be best conceptualised in their influencing the shaping of the goals. For reasons of simplicity, I will assume here that their motivating role in the problem-solving framework is captured in very general terms – they help increase the utility of the subject. I will thus abstract from the rich texture of human emotions and summarise their influence in the problem-solving procedure as inducing the subject to further her welfare, to better her condition or, in more technical terms, to increase her utility.[29] With respect to cognition, the approach that I suggest views the cognitive process as a complex mechanism that actively interprets and, at the same time, classifies the varied signals received by the senses. The mind classifies the experiences from the physical environment as well as those from the sociocultural environment. A wide variety of mental representations have been offered as cognitive models to describe mental operations that are of interest, but I will focus here on the *pragmatic notion of mental models*. Mental models gradually evolve during our cognitive development to organise our perceptions and keep track of our memories. As flexible cognitive structures, they are typically formed in pragmatic response to a problem situation, in order to orient the organism successfully in its environment.

One plausible way to account for mental representations is, thus, by way of mental models, which are coherent but transitory sets of cognitive rules; in other words, the classification of events and their interpretation in the light of a current problem occurs according to cognitive rules of the general "IF … THEN" type.[30] The mental model can be best understood as the final prediction that the mind makes or as an expectation that it has regarding the environment before getting feedback from it.[31] Depending on whether the expectation formed is validated by the environmental feedback, the mental model can be revised, refined or rejected altogether. Learning is the complex modification of the mental models according to the feedback received from the environment. However, the formation of mental models and the testing of solutions to problems in the environment do not necessarily lead to success – the fundamentally *hypothetical character of mental models* must be emphasised. Learning is an evolutionary process of trial and error,[32] and failure to solve a problem induces the search for a new solution.

A distinction between *old* and *new* problems is very useful in this context. Every time an individual is confronted with a problem situation, the human mind actively interprets and classifies it. When respective classes are available from earlier trial experiences under which the current messages of the environment can be classified, we can call the problem situation an "old problem." If the current problem is identified as a familiar one – in the sense that it can be classified in an existing class – then the appropriate solution will be applied automatically. A series of past successful solutions to the same problem has created what we call a *routine*. The main characteristic of a routine is that it is employed to solve a problem without any prior reflection, that is, without conscious effort and the mobilisation of cognitive resources. The use of routines is thus a function that does not require consciousness.[33] Routines employed unconsciously give rise to rule-following behaviour; and the crucial characteristic is that since this behaviour was often followed in the past, it has become standardised. This allows the limited cognitive faculties of the mind to be used thriftily; consciousness needs to be concerned with problems that are difficult to solve or with new problems.[34]

Whenever a solution to an old problem no longer works, possibly because of a change in the environment or whenever a problem situation, when compared to past problem situations stored in memory, cannot be classified under any familiar class, we speak of a "new problem." Naturally, all the old problems of an individual were new at some point in their history. There are two sequential kinds of response to new problems. The first response is quasi-automatic, and it involves the triggering of inferential strategies, that is, processes of inference. Cases of inference such as the description and characterisation of events, the detection of covariation among events, the use of causal inference, prediction and theory testing are normally solved by lay persons with the aid of judgmental strategies or heuristics.[35] Heuristics are to be understood as general strategies that provide quick solutions with little effort. They are, of course, also fallible.[36] And they can be used to solve both theoretical and practical problems in non-algorithmic ways.[37] Analogy is a very common and most powerful heuristic: if one has found a satisfactory solution to a problem in one domain, then an analogical transfer may lead to an equally successful solution in another domain.[38] If one knows how to ride a bicycle and she sits for the first time on a motorcycle, she will try to balance the same way as when she sat on a bicycle.

Hence, *the first step that one takes to solve a new problem, whether theoretical or practical, is to employ inferential strategies of a heuristic nature.* They are the first, intuitive response to any problem that is interpreted as a new one. There are two cases to be distinguished here: either the inference is successful and accordingly it will be strengthened and applied anew in the future or it leads to failure. Cognitive resources are mobilised in the form of attention, and the second stage of the problem-solving process ensues.

This is a deliberation process involving a mental probing of alternatives and a *choice* of the one alternative that is expected to best solve the new problem.[39] It is a conscious process, in contrast to the case in which old problems are solved by routines. The fundamental characteristic of this process is the *creation of alternatives* ex nihilo by the mind. In an imaginative process, the mind devises *new* alternatives.[40] *Creativity* is the property of the mind that is exemplified when working out new alternatives in attempting to solve a new problem. Considering one or more alternatives from the range of the ready-made alternatives sometimes available in the environment also remains a possibility. After the creation of the new alternatives, and often after the consideration of ready-made alternatives from the environment, an alternative is selected which seems to be the most appropriate to solve the problem. Like all other solutions, this is an entirely *conjectural* one, which cannot guarantee success. If the trial proves to be unsuccessful, then a new problem arises, and a new choice process is triggered. If the chosen alternative turns out to be successful in the environment, then it will be reinforced, and the next time the same problem arises, this solution will be reapplied. After it has been employed successfully many times, the solution will be standardised and will become a routine. In other words, the

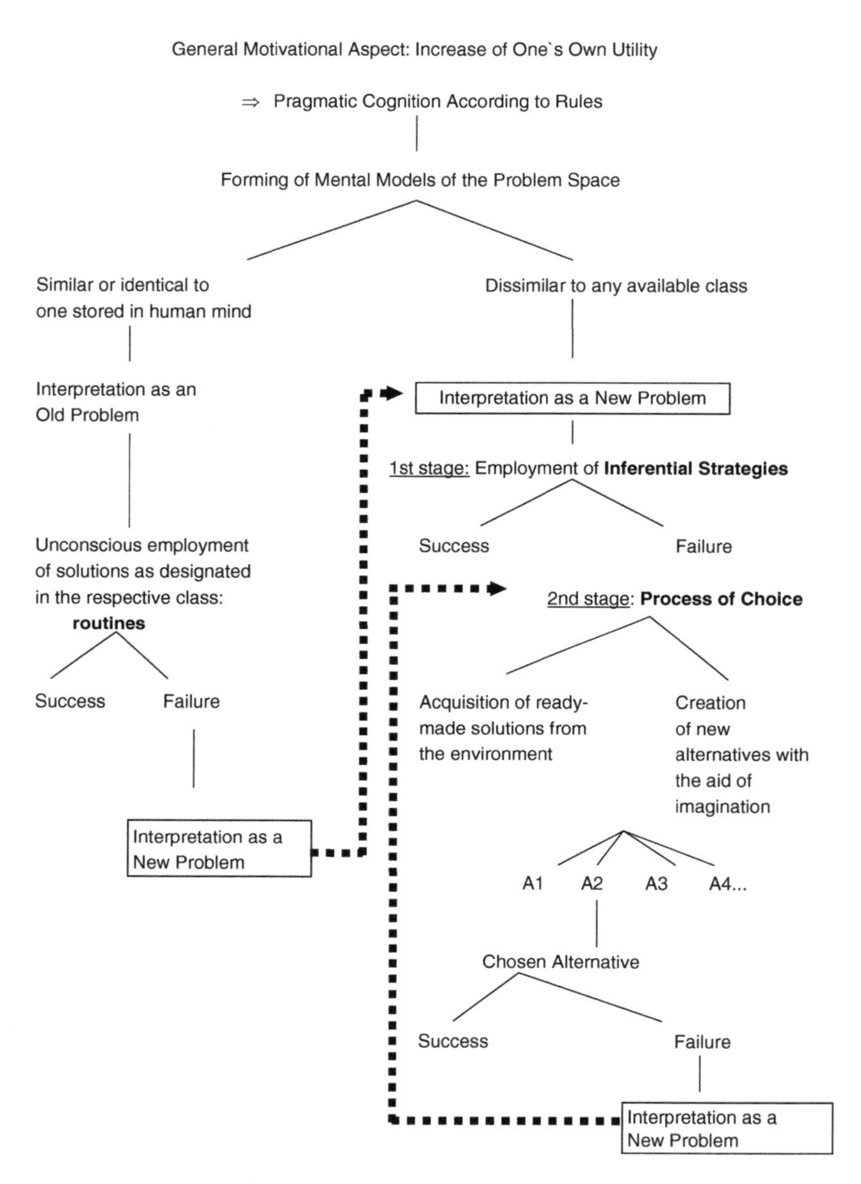

FIGURE 1 The problem-solving framework

mind will avail itself of a ready-made solution to the problem, and thus every time it is confronted with the problem, it will be classified as an old one and dealt with unconsciously.

This problem-solving framework portrays a scientifically informed account of such phenomena as use of intuitive heuristics, deliberation, choice and rule-following that constitute the first step towards the constitution of the "domain of the normative" and is summarised in Figure 1. When we are confronted with situations classified by our minds as *old problems*, unconscious routines are employed and *rule-following* is observable at the behavioural level. When we are confronted with situations classified by our mind as *new problems*, at a first stage *inferential strategies* are unconsciously employed and if they turn out to be unsuccessful, *creative choices* take place, fundamentally unpredictable from an observer's perspective. The successful solution of new problems leads to their reapplication and after a period of time to a *new routinisation*; and this process goes on and on.

The *appropriateness* of a solution to a given problem is what makes a specific solution successful in the circumstances, but up to this point there is no evaluation about whether it is *right* or *wrong*. Such an evaluation presupposes that the process of norm emergence proper has already taken place. How this occurs involves a second analytical step, which will be introduced in Section 3.3.

3.3 THE EMERGENCE OF NORMS

Moving from individual problem-solving to the interaction between individuals in different settings is the next analytical step. Individuals in any given environment continually communicate with other individuals while trying to solve their problems. The direct result of this communication, mimetic and verbal, is the formation of *shared mental models*,[41] which provide the shared framework for a common interpretation[42] and evaluation of the environment. Collective learning takes place along with a parallel emotional adaptation in an evolutionary process unfolding in historical time. As collective learning takes place at the group level, the problem-solving capacity of the society evolves and is transmitted. Diverse tools and innovations can be invented that further enhance or inhibit the processes of individual and collective problem-solving. A very common one is the invention of language, that accelerates the possibility of communication and substantially facilitates the collective learning process.

In this process of social interaction, shared mental representations and shared evaluations are created, mainly when two mechanisms are at work: They may emerge either deliberately or spontaneously, that is, either as a product of collective choice or as a product of a spontaneous process of social interaction. The latter mechanism was brought to light for the first time by the Scottish moral philosophers and is encapsulated in Adam Ferguson's observation that "nations stumble upon establishments which are indeed the result of human action, but not the execution of any human design."[43] Independently of the exact features of these mechanisms and their specification in different contexts, the main point is that *shared normative structures* always exist as the outcome of the deliberate design or spontaneous interaction among individuals. They are constituted by shared solutions to inter-individual problems, either as the outcome of a collective interpretation of a situation as a problem *tackled collectively* or as the outcome of an individual *first inventing* a standardised solution to an individual problem, and adopting and following a rule, which is later imitated by others in the group so that the rule gets diffused. The development of solutions to *shared problems* in a process of social interaction of whatever specific type is the sufficient condition for their evaluation as *right* or *wrong* ones. Appropriate solutions to individual problems can be *right* or *wrong* solutions. Their *rightness* or *wrongness* can be defended vis-à-vis other individuals in diverse ways – the exertion of physical force being the default strategy in the vast majority of cases in the recorded human history.[44] But it is also possible to defend their rightness or wrongness by the *use of reasons* in a communicative process.

There are myriads of shared problems and of problem solutions invented and adopted in human life. Whenever a problem is perceived and a solution to it is worked out, cognitive and evaluative moments co-exist. The nervous system of our evolved brains triggers mental representations of the environment that it evaluates as important for the attainment of the current goals of the organism. To the extent that language exists as a means of communication, it is possible to share mental representations of theoretical problems and their solutions. The mental representations of practical problems and their solution can be only shared via direct imitation, as I have stated previously. When language is available, a communication on the rightness or wrongness of the solutions both to theoretical and to practical problems among group members can take place. This is not some form of ideal communication among individuals, but communication taking place in historical time. Sometimes it involves reason giving. When the habit of recording ideas prevails, that is, of externalising the process of oral commentary and events, then more complex ideas can be placed in the public arena, in an external medium, where they can undergo refinement, extending the life-span of single individuals.[45] Deliberations about rightness and wrongness can then take more intricate forms, also written ones.

According to my naturalistic account normative structures emerge in processes of social interaction and are the outcome of problem-solving activities, involving creative choice and rule-following. The *content* of the norms depends on the exact structure of the social setting. One can classify stylised situations (e.g., but not necessarily, by using game theory) and analyse what kinds of shared normative structures emerge during the interaction among human problem solvers.[46] In different domains, different kinds of norms will emerge because the problems are usually (though by no means always) different: in science, in politics, in markets, in the family, etc. There is nothing mysterious in such processes: individual actors try to create and adopt those solutions and follow those rules that help them adapt to their current social environment.

3.4 REASON

Problems are very different from facts, and the solution of a problem cannot be equated with accounting for a fact. This does not preclude the possibility of a great range of problems being facts or states of affairs, but the class of problems includes much more than states of affairs. A problem does not need to accurately depict a state of affairs to be a problem; all that is needed is only that it be *thought* to be a state of affairs by one or more individuals. In other words, many problems are literally counterfactual: the problem of whether angels are male or female is of this type, as is the problem of describing the exact properties of a socialist utopia. Many facts about the world do not present themselves as problems simply because they are unknown. As Laudan observes, "a fact becomes a problem when it is treated and recognized as such; facts, on the other hand, are facts, whether they are ever recognized. The only kind of facts which can possibly count as problems are *known* facts."[47] Finally, problems can be transformed with the passage of time or they can even cease to be regarded as problems, whereas facts can never undergo transformation.[48]

My main intention here has been to show that the use of the problem-solving framework offers a plausible way to consistently and sufficiently accommodate the normative dimension while taking into account the deliverances of the sciences. I have mainly referred to social norms, that is, informal institutions. Normative rules enforced by the state, that is, formal institutions, are also a constitutive part of the institutional matrix. The institutional framework of a society need not and cannot be assumed to be "given." Rather, it has emerged in a historical process of individual and collective trial and error; and any institutional framework is only behaviourally relevant because it is anchored in individual minds. The upshot of my discussion is that normativity emerges in social groups when shared behavioural regularities anchored as routinised problem solutions in the minds of the members of the group are diffused and the deviators sufficiently sanctioned. All of these shared behavioural regularities always start as individual regularities.

The "domain of the normative" is constituted by a plurality and diversity of structures, that is, normative filters. These structures can be *very general*, particularly when they have been applied in many different contexts in a successful way and thus have been shown to constitute appropriate solutions to a great range of problems in diverse areas. Such structures are what we call *values* and constitute fixed encapsulations of normative resources applicable in many different situations. These can be epistemic values, moral values, political values, aesthetic values, etc. There is, naturally, value pluralism. Values providing the fixation of normative resources can themselves be used to evaluate normative rules, which are more context specific and transient in nature – this axiological level is the highest level in the hierarchy of *problems*, *rules* and *values*.

Reason is the general, specifically human faculty to *coordinate* all problem-solving activities of the human mind, to *create* and appropriately *evaluate* the solutions and to *embed* them in the natural and social environment. It manifests itself in creative choices, rule-following, and the capacity of learning by trial and error from the environment. Reason as a general faculty reveals itself in the various domains of interaction of the mind with its natural and social environment in different kinds of *rationality*: scientific, political, economic, etc. It is always context-bound but never entirely context-limited. The immanence of reason is never complete, and its more crucial feature consists in its possibility of transcending the given. The transcendence is piecemeal, creative, and critical.

Reason is our imperfect shield against relativism; how this shield can protect us in the case of science will be worked out in the following chapters.

4 The Informal Institutions of Science

The performance of science in relation to other forms of inquiry is preeminent. This is due to the fact that progress in knowledge has turned out to be possible in the course of history. Modern science is a constituent part of culture that also includes other systems such as art, religion and law. It is embedded, as are all such systems, in a social context without which it could not exist. Its characteristic form depends on a distinct institutional arrangement that channels its operation and allows its steering in a specific way (Jarvie, 2001).

Scientific progress has many facets and can be conceptualized in different ways, for example, in terms of problem-solving, of truthlikeness or of growth of knowledge. My main claim is that an institutional framework appropriately channelling competition and criticism is the crucial factor determining the direction and rate of scientific progress, independently of how one might wish to conceptualize scientific progress itself. My main intention is to narrow the divide between traditional philosophy of science that elaborates on the standards defining science as a *truth-seeking enterprise* and the sociological, economic and political outlook on science that emphasizes the private interests motivating scientists and the subsequent *contingent nature of the enterprise*. My aim is to show that although science is a social enterprise taking place in historical time and thus is of a contingent nature, it *can* and in fact *does* lead to genuine scientific progress – contrary to the claims of certain sociologists of science and other relativists who standardly stress its social nature but deny its progressive character, following the lead of Bloor (1976/1991), Latour and Woolgar (1979/1986) or Rorty (1979).

4.1 INSTITUTIONS: EMERGENCE AND EVOLUTION

"Institutions keep society from falling apart, provided that there is something that keeps institutions from falling apart" (Elster, 1989b, p. 147). This phrase by Jon Elster nicely summarizes the relevance of institutions for society.[1] Institutions are normative social rules, that is the rules of the game in a society, enforced either through the coercive power of the state or other enforcement agencies that shape human interaction (Mantzavinos, 2001). They constitute normative patterns of behaviour that provide solutions to problems of coordination and cooperation in society in virtue of offering a quasi-permanent platform of conflict resolution. Institutions as the rules of the socioeconomic game define what kind of strategies and which action parameters can be employed by agents in their activities.

Institutions must be distinguished from organizations (North, 1990). Institutions are the *rules of the game*; organizations are *corporate actors*, that is, groups of individuals bound by some rules designed to achieve a common objective (Coleman, 1990). They can be research organizations such as universities, political organizations such as political parties or economic organizations such as corporations. Organizations and individuals are the *players in the game*. When individuals and organizations interact, they are attending to the general normative rules which we call institutions, that is, the rules of the game, that constrain their behaviour.

After defining institutions and providing the distinction between institutions and organizations, we can ask the most fundamental question of the theory of institutions: why do institutions exist? One has used different approaches to answer this question, but it is the individualistic approach that I will follow here, according to which there are two broad classes of causes that can explain the existence of institutions. The first class includes the causes that have to do with the motivational structure of *Homo sapiens* and the second with the cognitive one. If one starts with the common hypothesis about motivation, that is that individuals strive to better their condition by the means available to them, formalized in the idea of increasing her own utility, then an interindividual conflict is bound to arise. The patterns of such conflicts are distinct and observable from an observer's point of view and are systematized and formalized by contemporary game theory (e.g., Dixit, Skeath, and Reiley, 2015; Guala, 2016), such as the well-known prisoner's dilemma, the coordination game, the game of trust and many more. Such settings are "social" not in the sense that the agents are consciously aware of them, but in the sense that they are identifiable by the social scientist. The most fundamental raison d' être of institutions is, thus, that they provide workable solutions to social problems, most of which of a conflictual character. Institutions help partially overcome the Hobbesian problem: the life of man in a society *without institutions* would be, in the words of Hobbes (1651/1991, p. 89), "solitary, poore, nasty, brutish and short."

But why do people adopt institutions, that is, social normative *rules*, rather than deciding *each time anew* on how to solve a social conflict? Since every problem situation has unique characteristics, why not solve social problems ad hoc? The second class of causes for the existence of institutions provides an answer to this question. These causes have to do with the cognitive structure of humans. Cognitive capacities are limited and are mobilized only when humans are confronted with "new problems" in their environment; they follow routines when they classify the problem situations as familiar ones (Mantzavinos, 2001; Mantzavinos, North, and Shariq, 2004). This distinction is rooted in the limited computational capacity of human beings (Simon, 1983; Gigerenzer, 2008, 2021). In a genuinely uncertain environment, that is an environment characteristic of a non-ergodic world (North, 2005, ch. 1), the mind must be freed up from unnecessary operations, so that the problems in such an environment can be tackled at all and dealt with adequately. In other words, a huge number of mental processes becomes automated and do not take place in the light of consciousness and this is the basic cause of showing a routinized behaviour.[2]

Our limited cognitive capacity makes our environment appear rather complicated to us and in need of simplification in order to be mastered (Heiner, 1993) – this is what we mean when we say that our environment is complex. This refers to both our natural and our social environment, the latter being the focus here. Rules in general, as Hayek (1976/1982, p. 8) put it, "are a device for coping with our constitutional ignorance," they are the "device we have learned to use because our reason is insufficient to master the full detail of complex reality" (Hayek, 1960, p. 66). Institutions are our devices to deal with *recurrent social problems* arising in situations where self-interested individuals interact.

A very fruitful and widely used criterion for distinguishing among different types of institutions is according to their enforcement agency. An institution is not simply a social rule shared by individuals (and organizations) but also the enforcement characteristics of it. One can classify institutions according to this criterion as shown in Figure 2.

It is impossible to provide a detailed analysis of the emergence, working properties and enforcement of the different types of institutions here.[3] I would only like to highlight two *general mechanisms* of the emergence of institutions: they emerge either deliberately or spontaneously, that is, either as a *product of collective decisions* or as a product of a *spontaneous process of social interaction*. The deliberate emergence of institutions has long been the object of inquiry since their explanation is relatively simple – they are explained

<table>
<tr><td rowspan="3">Informal institutions</td><td>Conventions</td><td>Self-policing</td></tr>
<tr><td>Moral rules</td><td>First party</td></tr>
<tr><td>Social norms</td><td>Third party: other individuals in the group</td></tr>
<tr><td>Formal institutions</td><td>Law</td><td>Third party: state</td></tr>
</table>

FIGURE 2 The classification of institutions

exclusively by intentional action aimed at establishing them (Knight, 1992). The spontaneous emergence of institutions (Hayek, 1973/1982, p. 37) as originally conceptualized by the Scottish moral philosophers requires more sophisticated explanatory patterns (e.g., Bicchieri, 2005, 2016). Although the enforcement agency and the specific enforcement mechanism is different for each category of informal institutions, there is a common element to each type of informal institution, that is, conventions, moral rules and social norms: *they all emerge as the unintended outcome of human action*. The emergence of informal institutions is, thus, the outcome of a process that is not under the conscious control of any individual mind. Law or the class of the social rules that I have called formal institutions are, on the contrary, products of collective decisions. *The state as an organization* creates law either by providing by means of suitable adaptation existing informal institutions with sanctions or by constructing, by the conscious decision of its organs, altogether novel legal rules. Modern public choice theory provides explanations of how collective decisions lead to the emergence of institutions in the social realm (Mueller, 2003). In a nutshell, during the political process individuals and organizations aiming at furthering their interests succeed in a greater or lesser degree – while using the power that they have – in influencing the collective decision-making procedures that lead to the creation of legal rules. What is called "political power," however, is a factor that is very difficult to identify. Therefore, contemporary positive political theory often uses the proxy of "resources" in order to determine the behaviour of the players in the political game. There are usually three kinds of resources that players use in their endeavours, that is, economic, political and ideological. The extent of the bargaining power of the players is determined by the degree of their availability. Consequently, the resources decisively influence the political process that in the end generates the formal institutions (Moe, 2005).

It is the case, thus, that the mechanisms for the emergence of informal and formal institutions are distinct. The informal institutions emerge endogenously from within society as the unintended results of human action during an invisible-hand process, whereas the formal institutions emerge in a way exogenously, in the sense that they are the outcome of the political process driven by the collective decisions of agents availing themselves of resources: political, economic and ideological. There is therefore no necessity that informal and formal institutions complement each other in such a way that a workable social order is produced or that scientific progress takes place.

4.2 SCIENCE AS A GROWN ORDER

Scientific activity is undertaken by imperfect biological organisms with a limited cognitive capacity in interaction with artefacts in a specific social context. The scientific enterprise is a social process (Hull, 1988), and it consists of the attempt of the participants in this process to provide answers to puzzles and solutions to theoretical problems (Mantzavinos, 2013, 2016). The scientific enterprise is embedded in the institutional framework of the society consisting of informal and formal institutions. What we call "science" is not a means towards the accomplishment of anything. It is, instead, the institutional embodiment of the processes of constructing and criticizing solutions to theoretical problems that are entered into by individuals in their several abilities and skills. Individuals are observed to cooperate with one another, to compete with one another, to devise representational vehicles for solving problems and to experiment and criticize one another. The network of relationships that emerges and evolves out of this process is called "science." It is a setting, an arena, in which scientists attempt to accomplish their own purposes, whatever these may be.

The talk of "the aim of science" is not simply an inexpedient abstraction; it is a seriously misleading oversimplification. Only an agent can have an aim. This is evident in the case of an individual agent. In the case of a group, one can still plausibly defend the position that it can be viewed as an agent and, thus, also have an aim (List and Pettit, 2011, ch. 3). Organizations, for example, whose internal life is regulated by a set of organizational rules (to which naturally only the members of the organization are bound and not all members of a society), might also be plausibly conceived as following an aim – this is the aim, which is specified by the internal organizational rules: provision of teaching in a school organization, for example, or provision of research in a research institute or profit seeking in a corporation. But the order of activities that we call "science" is not a deliberate arrangement made by somebody. It is a grown order exhibiting orderly structures, which are the product of the action of many individuals but not the product of a human design. It constitutes an arena of activities which has not been made deliberately – therefore, it cannot legitimately be said to have a particular aim.

Some well-known views of the aim of science like van Fraassen's view that it consists in the empirical adequacy of its theories (1980, p. 12) or Potochnik's view that it consists in providing understanding (2017, p. 93ff.) or Bird's view that it consists in knowledge (2019, p. 173ff.) are untenable, because they conceptualize science as if science were a *deliberate* collective enterprise.[4] The only plausible question, on the contrary, is whether and how the *diverse aims* of individual scientists and scientific organizations produce outcomes in a process of social interaction that are appraised positively with reference to diverse values. This will be made clearer in Chapters 5 and 6.

If one endorses the conceptualization of science as a grown order that I propose, the main issue becomes what kind of rules guide the epistemic problem-solving activities of the scientists in the arena of science and how, if at all, scientific progress takes place. Karl Popper (1944/1957, 1945/2002) was the first to elaborate an institutional theory of scientific progress. He suggested as a starting point for such a theory "to try to imagine *conditions under which progress would be arrested*. [...] How could we arrest scientific and industrial progress? By closing down, or by controlling scientific periodicals and other means of discussion, by suppressing scientific congresses and conferences, by suppressing Universities and other schools, by suppressing books, the printing press, writing, and, in the end, speaking. All these things that indeed might be suppressed (or controlled) are social institutions. Language is a social institution without which scientific progress is unthinkable, since without it there can be neither science nor a growing and progressive tradition. Writing is a social institution, and so are the organizations for printing and publishing and all the other institutional instruments of scientific method. Scientific method itself has social aspects. Science, and more especially scientific progress, are the results not of isolated efforts but of the *free competition of thought*. For science needs ever more competition between hypotheses and ever more rigorous tests. And the competing hypotheses need personal representation, as it were: they need advocates, they need a jury, and even a public. This personal representation must be institutionally organized if we wish to ensure that it works. And these institutions have to be paid for, and protected by law. Ultimately, progress depends very largely on political factors; on political institutions that safeguard the freedom of thought: on democracy" (Popper, 1944/1957, p. 154f.)

In his "Republic of Science," Polanyi has presented science as an arena that exemplifies the principle of spontaneous coordination by mutual adjustment. It is a field in which "self-coordination of independent initiatives leads to a joint result which is unpremeditated by any of those who bring it about" (Polanyi, 1962, p. 55). The metaphor of the invisible hand can be used here, in order to describe the mechanisms that are at work transforming the diverse private interests into a specific order, aggregating the dispersed individual activities into the patterned outcome (Ullmann-Margalit, 1978, p. 267f.).

Building on the work of these pioneers, I would like to elaborate in the forthcoming sections of this chapter and in Chapters 5 and 6 on the informal institutions of science, the formal institutions of science and how scientific competition and collaboration unfold within these rules, so that the view of science that I would like to convey becomes clearer.

4.3 THE INFORMAL INSTITUTIONS OF SCIENCE: THE SCIENTIFIC METHOD

According to the general account of the emergence and evolution of institutions that I have provided in Section 4.1, institutions are normative social rules, that is the rules of the game in a society enforced either through the coercive power of the state or other enforcement agencies that shape human interaction. Institutions regulate all aspects of human life solving conflicts and coordinating activities of the social actors, that is individuals and organizations (Picavet, 2020). Institutions regulate the political life, the economic life, the religious life and also the scientific life of the members of a society. In these distinct domains of social interaction typically different kinds of rules have emerged, so that the rules that structure politics are different from the rules that structure markets and from the rules that structure religious activities. There are of course in many cases similarities among the rules, but the very distinction between domains is greatly due to the different institutions that regulate each of them. The social rule of non-violent succession of governments along with a series of other domain-specific rules regulate democratic politics, for example. The social rule of private property rights on the means of production with a series of other domain-specific rules regulate competitive markets, and so on. All these domain-specific rules determine to a great degree the way that the specific activities of the respective agents will unfold; they structure the respective game: the political game, the economic game and the scientific game.

The domain-specific rules that regulate science have evolved in a very long historical process. The need to survive in an environment inhabited by other species, the practical need to improve human life and the wondering about the natural world were among the many epistemic interests that human beings had since the dawn of civilization. There has always been a diversity of interests motivating the specific epistemic activities in different phases of history in different parts of the world. In the case of the Western world, the transition from *μύθος* to *λόγος* (Snell, 1946/2009, ch. XI) has taken place in Ionia where the first Presocratic philosophers made the decisive move from dogmatically trusting narratives handed down by older generations to making use of a broader set of cognitive abilities and intellectual faculties in order to satisfy their epistemic interests. That this move activated a broader set of cognitive abilities including both theoretical contemplation *and* practical activities is nicely encapsulated in how such an innovator came to be characterized by his contemporaries: he has been referred to as a philo*sopher* (φιλόσοφος in Greek) – a friend of wisdom – rather than as a phil*epistemon* (φιλεπιστήμων in Greek) – a friend of knowledge. Wisdom requires the successful use of practical skills along with the theoretical ability of contemplation.

The long historical process from ancient philosophy to modern science (Fara, 2009) is characterized by the evolution of the rules used to engage in epistemic activities. New rules have been invented by innovators to represent reality and to offer orientation in the world, they were often adopted by imitators when they were deemed to be successful, and they were periodically diffused within broader epistemic communities. These rules have always been diverse, suited to serving different epistemic problem-solving activities.[5] The majority of the ever-occurring innovation-imitation cycles throughout history usually affected an *incremental change* of the rules guiding epistemic activities. The diffusion of successful rules in the population of epistemic agents is mostly incremental since this is a spontaneous process of interaction requiring the formation of shared mental models and the parallel process of emotional adaptation when a new set of rules is adopted – these processes simply take a long time to be completed, even in the rare cases where there is no conflict of interest in adopting new rules and abandoning the old ones. But there are also cases, of course, where more rapid changes occur, sometimes taking even a revolutionary character. As many historians of science seem to agree, the Scientific Revolution was such a case.[6]

What was revolutionary in the Scientific Revolution, which gave rise to modern science? First of all, the quality of the epistemic products themselves was revolutionary – just think of the marvellous theories of Kepler, Newton or Harvey, theories that were more empirically accurate, simpler and closer to a true description of aspects of the natural world than everything else invented in the past.[7] But what is even more important than that, what was also revolutionary was the invention and widespread use of a set of rules that have come to be summarized as the *Scientific Method*.[8]

Before the Scientific Revolution, in different parts of the world and in different historical epochs, bold theoretical conceptions of the inner workings of nature have been invented and diffused. In ancient China after an initial competition among diverse ideas, a synthesis has emerged and prevailed under Han, the first stable imperial dynasty. An important crystallizing point of the ancient Chinese thought was that the maintenance of a stable social order could take place, if human nature could reflect the harmony of the cosmos. The leading idea was that the multiplicity of natural phenomena should be viewed under the lens of their interdependence, and this mode of thinking was called "correlative." As Needham (1978, p. 165) stresses: "For the ancient Chinese, things were connected rather than caused [...] The universe is a vast organism, with now one component, now another, taking the lead at any one time, with all the parts co-operating in a mutual service [...] In such a system as this, causality is not like a chain of events [...] [rather] succession was subordinated to [...] interdependence."

In Athens, all four philosophical schools, the Platonic Academy, the Lyceum of Aristotle, the Stoa and the Epicurean Garden, have developed quite sophisticated theoretical structures in order to shed light on the cosmos. All the approaches propagated were top-down approaches, trying to deduce from a few first principles a great range of natural phenomena. Though observations did play some role, most importantly and systematically by Aristotle regarding the living organisms, the conceptions remained fundamentally theoretical and the debates between the schools remained inconclusive, since every school defended the first principles that it took to be fundamental and from which everything else should be deduced. Plato introduced the distinction between the imperfect world that we observe with our senses and a perfect world of Ideal Forms of which the objects that we perceive with our senses are a poor reflection. A genuine knowledge of the natural world is possible only when grasping the Ideal Forms. Aristotle introduced a distinction between two kinds of Being: Potential and Actual and his theoretical conception consisted in working out how change could be theoretically grasped as the growth of what something carries inside itself as a possibility into a state of actuality. Epicurus's atomist doctrine introduced an infinite number of minuscule fragments of Being that were called 'atoms', that is, indivisible, and the natural world is constituted as an endless reconfiguration of these atoms. Finally, the Stoa stressed the unbreakable continuity of the natural entities rather than their physical separateness. Pneuma is the intermediate substance that permeates the entire cosmos and induces this continuity.

It should be stressed that these theoretical conceptions, admirably ingenious and creative as they have been, could barely come into fruitful contact. Although there has been a debate among the proponents of these imaginative constructions, the debate was of a dogmatic character since the first principles of the respective schools were not supposed to be amenable to revision (Cohen, 2015, p. 12f.). There was no consensus about any acceptable way that could lead to the revision or abandonment of a theoretical position; there was no consensus, in other words, with respect to a possible arbiter of the disputes. Such a consensus about an acceptable arbiter has emerged during the Scientific Revolution, and this was the major methodological revolution. Strevens (2020, p. 93) describes the content of the main methodological rule that has become accepted by the natural philosophers active in the Scientific Revolution (the term "scientist" was introduced much later, in 1833, by Whewell): "What the rule says is simple enough: it directs scientists to resolve their difference of opinion *conducting empirical tests* rather than by shouting or fighting or philosophizing or marrying or calling on a higher power. That is all; it makes no attempt to interpret the evidence, to decide winners and losers. Indeed, its function is not so much to resolve the dispute as to prolong it. This perpetuation of the dramatic conflict for its own sake is the essence of the scientific method." [My emphasis, C.M.]

It might be a fruitful simplification to speak about a single methodological rule in this context. I propose, however, to consider a broader set of rules that the natural philosophers have adopted in the accelerated collective learning process that took place during the Scientific Revolution. It was a whole set of informal institutions that came to be adopted in a widely shared consensus, which was the key to a successful understanding of natural phenomena. This set specified for the first-time stricter demands on *how evidence should bear on theory* that included quantifiable standards of evidence as well as a new source of evidence: experimentation. A peaceful dialogue is certainly a start for a civilized intellectual exchange, but it is hardly sufficient to bring about knowledge about natural phenomena and to set in motion successful progressive epistemic problem-solving activities; this will only be enabled when imaginative theoretical constructions are appropriately fitted with evidence provided from diverse sources.

According to the general classification of informal institutions that I have offered at the beginning of this chapter, it is useful to briefly specify how the way that evidence should bear on theory has been incrementally crystallized during the scientific revolution in concrete (1) *scientific conventions*, (2) *moral rules that scientists adopted* and (3) *scientific techniques*. Starting with the scientific conventions, recall that these rules are self-policing, because there is no genuine conflict of interest between individuals following them.[9] Among the many conventions, the one that should be particularly stressed, is the *use of the printing press* in circulating ideas, discoveries and news. This was a fifteenth-century invention whose impact grew through the seventeenth century constituting a novel convention that (almost) everyone was interested to adopt, because (almost) everyone had an interest in communicating his ideas quickly and effectively. The widespread use of books and other printed material constituted a huge change in the mode of communication between natural philosophers during the scientific revolution. A well-known example is Vesalius's *De Humani Corporis Fabrica Libri Septem* (On the Fabric of the Human Body), which was published in 1543 (the same year as was published Copernicus's *De Revolutionibus Orbium Coelestium*). The drawings of the human body contained in the Fabrica were the outcome of the collaboration of Vesalius with Jan Stefan van Calcar from the workshop of Titian, who constructed the woodblock engravings that were then printed in the atlas. These anatomical drawings and illustrations constituted a productive synthesis of art and science and reflected a naturalism in the depiction of the human anatomy that differed radically from the medieval drawings. Vesalius's work, whose actual printing he has personally supervised in Basel, had the effect that the drawings could be multiplied with accuracy and made easily accessible to all interested scholars. This became the canonical anatomical textbook for a very long period of time, its last printing taking place in 1782.

Coming now to the second type of informal institutions according to my categorization at the beginning of this chapter, typical *moral rules* include "keep promises," "do not cheat," "tell the truth," etc. Apart from the rules of general morality, what is more important in the context of the Scientific Revolution were the specific moral rules necessary in order for the specific epistemic task of bringing theory and evidence together to be accomplished successfully. As Wootton stresses (2015, p. 422): "For when scientists began to make assessments regarding the reliability of evidence they were required to exercise what they all called 'judgment' (for example Locke: 'Knowledge being to be had only of visible certain Truth, *Error* is not Fault of our Knowledge, but a Mistake of our Judgment giving Assent to that, which is not true[10]'). And exercising judgment requires a specific set of virtues, the virtues you would hope to find in a jury of your peers: impartiality, assiduity, sincerity. We find these virtues everywhere in discussions of Evidence-Indices [...]." It seems, thus, that specific moral rules such as *impartiality*, *assiduity* and *sincerity* were found to be exceptionally important in the epistemic community of the natural philosophers who were willing to judge the quality of their theoretical inquiries into the natural world based solely on empirical evidence.

The third kind of informal institutions according to my classification encompasses all social norms that emerged and were diffused during the scientific revolution guiding the epistemic problem-solving activities of the natural philosophers of the time. To restate Strevens's point, the most important norm was that epistemic problem solutions should be primarily making use of empirical tests (2020, p. 98): "Scientific argument, unlike most other forms of disputation, has a valuable by-product, and that by-product is data. [...] [A]s long as the protagonists are practicing science, their arguments must be conducted by experimental means. All of their need to win, their determination to come out on top – all of that raw ambition that, on modern sociological view of science, would subvert any objective code of inquiry – is diverted into the performance of empirical tests. The rule thereby harnesses the oldest emotions to drive the extraordinary attention to process and detail that makes science the supreme discriminator and destroyer of false ideas." These social norms of science I would like to call "scientific techniques." I use the term "technique" for two reasons, first, in order to stress the practical aspect along with the theoretical and second, in order to reserve the term "scientific method" for the entire set of informal institutions that guide epistemic problem-solving. The most important scientific techniques along the one that has already been referred to by Strevens are the well-known ones from most historical accounts of the Scientific Revolution: the rules that describe the ways in which *measurements of natural phenomena* should be made; rules that determine the ways that *experiments should be conducted*; rules of *catalogizing and presenting the data*[11]; rules of *evaluating the data*, etc.

Summarizing my argument up to now, my claim is that there is indeed a Scientific Method. It is constituted by the informal institutions that have originally emerged and diffused during the Scientific Revolution, consisting of essentially three types of rules, the scientific conventions, the moral rules and the scientific techniques. The term should be understood as a *type*, with the different informal institutions of science being the tokens. The scientific method refers to the general way that theory and evidence is brought together by scientists, and this way is instantiated by the different rules that the scientists follow when they try to successfully accomplish this general task. The context within which this general task is performed changes due to many reasons. For example, when extending the scope of application of the task to more and more areas of research; when new artefacts are invented enabling new ways to fit theory and evidence; when new organizational structures are invented in which resources are pooled together enabling scientists to engage in novel forms of interaction, etc.

The claim that the informal institutions of science evolve over time is perfectly consistent with the claim that there is a scientific method – it is a type exemplified by multiple tokens, that is different informal institutions in different historical periods and in different parts of the world. As money can be exemplified in very different entities that serve the functions of facilitating exchange, measuring economic value and storing wealth, so the scientific method can be exemplified in different rules that serve the function of successfully bringing into contact theory and evidence. There is absolutely no inconsistency in claiming that the scientific method has been invented during the Scientific Revolution and that it is still at work today, even if the rules that have been originally invented and applied to natural phenomena have successively been applied to social phenomena, have been modified tremendously and have been refined to solve very intricate epistemic problems. The scientific method may take very different forms, but as long as *it consists in the techniques of bringing the products of our theoretical imagination in contact with empirical data by scientists following the scientific conventions of their time and the moral rules necessary for this kind of epistemic problem solving*, it remains a distinctive way of grasping the structure of the world.

The view of the scientific method that I defend here is opposite to a series of more simplistic views that attempt to pin down the method of science to one simple formula: there is no simple algorithmic rule whose mechanical application can provide a secure way to get access to the secrets of nature and society. It is rather a whole set of informal institutions, which, when followed by the participants in the arena that we call "science," offers the comparatively better way to represent the natural and social environment than all other ways that human beings have invented in the past. There is no guarantee that this is the best possible way to successfully solve epistemic problems, but it is the best among the available alternatives that we have been able to invent so far. Whether, rather than elaborating on the existing rules, we will be able to create rules of an entirely novel form in the future is something that is unknown to us today.

4.4 THE INFORMAL INSTITUTIONS OF SCIENCE: THEORY AND EVIDENCE

It is useful for our further discussion of the scientific method to distinguish between the context of discovery, the context of justification and the context of application.[12] Epistemic problems are conceived and formulated in the context of discovery; evidence is brought to bear on the generated problem solutions in the form of diverse representational entities such as theories, models or hypotheses in the context of justification; scientific outcomes, empirically warranted to a greater or lesser degree, are used to intervene in the natural and social world according to the goals of the users of these outcomes in the context of application.

I shall say more about the context of discovery and the context of application at the end of this chapter. For now I will focus on the context of justification – this is where the scientific method is at work. It is the context in which evidence in very diverse forms and from very diverse sources is used to test the validity of the theoretical constructions produced by scientific agents, both individuals and organizations. What is very important in this context and cannot be stressed enough is that what we call the interplay between theory and evidence is more specifically the interplay between *representational entities* and *different kinds of observations*, such as observational reports, experimental observations or reports from findings of excavations. As I will elaborate in Chapter 5, representation is a triadic relationship including three relata: the representation-bearer, the representational object and the interpretation in a cognitive agent. Representation is constituted by these three relata and the relations obtaining among them. It should be obvious that propositions do not exhaust the great variety of representational entities, that is, entities that bear representation. Therefore, the discussion in traditional epistemology that typically focusses on the properties of propositions and how they relate to other propositions and with evidence is extremely limited. There is a great variety of scientific products in the form of scientific representations: graphs, computer monitor displays, diagrams, scale models, mathematical models, etc. How these diverse representational entities can be assessed by the use of evidence is an extremely more complicated problematic that traditional epistemology dealing mainly with propositions does not do any justice to.[13]

These representational entities can often take the form of material models: think of the metal plates representing four bases along with rods arranged helically around a retort representing DNA. However, material models of this kind are constructed as representation-bearers, and not as technological artefacts. In other words, what is assessed in these cases are their *representational role* and *not the efficacy of their use* to accomplish some other goals in order to intervene in the natural environment – and this is a categorical distinction between science and technology,[14] between scientific constructions and technological artefacts.

These theoretical constructions possess properties: they can be true, simple, empirically accurate, fruitful, etc. My main claim is that all these properties are descriptive properties. There is nothing essentially evaluative about truth or simplicity or empirical accuracy. They are descriptive properties such as speed, toughness or thermal conductivity. These are distinctive descriptive properties of representational entities, since, trivially, not all entities can have all kinds of properties. I want to elaborate on this point with respect to the property of truth, which is of fundamental importance and then proceed by analogy to argue with respect to other descriptive properties of representational entities such as simplicity and empirical accuracy.

Suppose that you feel sick and you go to the doctor who formulates the hypothesis that you have cancer and orders that you do a series of medical tests. You decide not to proceed with the medical tests, because you do not want to find out the truth about your disease. You do not want to know whether you have cancer or not, because you want to go on living your life as you used to. In this case, you do not attach as much value to truth as you attach to health. In other cases, probably in the vast majority of all other cases, you would attach more value to truth, but not in this one, for whatever reason.

As this example shows, there is nothing constitutively normative about truth itself. Truth is a descriptive property of a representational entity, and it is we that attach value to it. In individual cases like the one in the example, it is an individual decision to attach or not to attach value to truth. In cases of social interaction like the case of scientific processes, we can either collectively decide to value truth or come to spontaneously adopt rules that attach value to truth.

It is one thing to claim that the discourse that leads to the production of judgments is normatively guided, through and through, a position that I endorse. It is another thing to claim that the content of all judgments is *constitutively* normative.[15] A non-naturalistic stance postulating a peculiar kind of norm constituting the truth content of a judgment cannot give a specific account of such a mystery. Besides, to claim that truth content is constitutively normative would have the absurd result that *all* our judgments are value judgments. Who is willing to accept this result?

When I say that truth is a descriptive property of representational entities, I mean truth in the sense of the long tradition of Western philosophy, that is, truth as correspondence with the facts.[16] This is the conception of truth that is used in everyday discourse, by the vast majority of the scientists and by many contemporary philosophers (though a series of other philosophers reject it).[17] It is common in standard epistemology to discuss the truth of propositions with respect to simple facts like in the case: "'The cat is on the mat' is true if and only if the cat is on the mat." The discussion is often about truth-makers, that is, what do facts consist in and how exactly they make a statement like 'The cat is on the mat' true. Although prima facie interesting, the air of generality that such discussions convey is to be resisted.[18] The implicit claim is that by conceptualizing our cognitive mental states as *beliefs*, our linguistic expressions as *statements* and by discussing how they *relate* to *states of affairs* in the world is an adequate way to deal with the issue. Although presenting the situation in such a way might be of some use for characterizing the truths that we are able to learn in our everyday life, it is a far cry from the situation that characterizes the much more ambitious scientific endeavours to disclose the secrets of nature and society.[19]

There is a long series of representational entities that are created by scientists, and there is a variety of descriptive properties that these may have. Mathematical models, for example, often make deliberate use of idealizations, so that from their very construction they cannot be true. They are false models, which can have other properties, like, for example, predictive success. In short, there is a diversity of scientific representations, which can have different descriptive properties other than truth: fruitfulness, simplicity, internal consistency, coherence with other representations, etc. All these properties are *epistemic* properties, simply because the entities are representational entities. A mathematical model cannot be unmarried, a statement concerning a natural constant cannot have a brother and a computer monitor display cannot be rocky.

Exactly as in the case of truth, it is we who attach value to the different epistemic properties of the representational entities that are produced during the scientific process. We attach value to the simplicity, fruitfulness, coherence or empirical accuracy of these entities. Representational entities and their properties are not bearers of value, *it is human beings who evaluate them*. As with the case of truth, there is nothing *constitutively* normative in these properties, which are all descriptive. Judgments such as "Model *x* is simple" and "Computer model display *z* is accurate" are descriptive judgments.

In sum, it is important to remember that scientific products are *epistemic* products, they have *epistemic* properties, and these properties can be described by descriptive judgments. Outcomes of other kinds of social processes, for example, an economic production process, are material products that have *non-epistemic* properties. A car, for example, has specific properties such as speed or a number of seats, but a car cannot be empirically accurate or true.

Now, the scientific method consists in the set of *scientific conventions*, *specific moral rules* and *scientific techniques* that enables scientists to bear different sources and pieces of evidence on their theoretical constructs. As I have repeatedly stated, these are normative rules that guide the activities of scientists in assessing the epistemic properties of the representational entities that have been created. The arena of science being a social domain is constituted by specific norms that the actors in this domain share. As scientific activities unfold, representational entities are constantly generated as the products of these activities. But they are not themselves normative. They just are.

Because of the complexity of the problems that scientists are often dealing with, it is very seldom the case that it is possible to produce descriptive judgments of the type 'The cat is on the mat' which correspond to the respective facts. *Markers of truth* are needed, and these markers of truth are the other descriptive properties that I have been discussing earlier: empirical accuracy, consistency, coherence, simplicity, fruitfulness, etc. These properties are distinct and can be described with the aid of diverse descriptive judgments depending on the nature of the specific representational entities under focus. It is not the case that any of the above properties is a sufficient or necessary condition for truth. To take an obvious example, a theory might not be simple, but still be true. It depends on the concrete context, if and how the different descriptive properties are related to truth, but it is often the case that their existence raises the probability of a theoretical construct being true.

Why is all that important in our context? Because it seems that there is a need for more clarity in the respective debates about the "value-free ideal" of science with respect to these matters. As I am discussing in the Excursus of this book, there is a great array of views ranging between the two extreme positions of defending a "value-free science" and of completely rejecting the "value-free ideal" proposing that non-epistemic values, that is, moral, political, emancipatory etc. should be used to make decisions with respect to the acceptance or rejection of scientific hypotheses. It should be clear that according to the general approach that I defend here, there *cannot be* a "value-free science" since science is a social process where norms emerge and are adopted in order to solve a great array of problems constitutive of scientific activity. The naturalistic account of normativity that I have presented in Chapter 3 describes how problem-solving activities are conducted by individuals, and how rule-following behaviour and choice are ubiquitous phenomena in the "domain of the normative." Scientists also in the context of justification follow normative rules and make choices when bringing evidence to bear on theoretical constructs. Since, as I have stated in that chapter, values encapsulate normative resources at the highest and most abstract level, they are prevalent in the activities of scientists in the context of justification. *All science is value-laden* since all scientists follow rules and make decisions on the basis of their general faculty of being endowed with reason. The outcomes of their activities, however, are not themselves normative.

To clarify my position with the help of an example that is often discussed in the literature (Douglas, 2000, 2009): when a scientific hypothesis is tested on the basis of a statistical significance test (say whether a specific chemical such as dioxin is dangerous for health), it is clear that stricter standards of statistical significance in comparison with control groups will reduce the number of false positives and increase the number of false negatives. This will make the chemical appear less dangerous than it actually is. If it is decided that laxer standards of statistical significance should be used, then the number of false positives will increase and the number of false negatives will decrease. It could seem that prima facie the warrant of the scientific hypothesis will depend on a previous decision of setting higher or lower standards of significance. It is indeed the case, that such decisions are value-laden, but there is *nothing normative in the statements declaring the outcomes of the respective tests.*[20] And all scientists are perfectly aware of what a margin of error means in each case. The decisions are scientific activities, the statements describing the tests are the outcomes of these activities; the first are value-guided, the second are descriptive judgments.

What kinds of rules are followed by scientists in the context of justification? I will only refer to the most general ones and so the list is not exhaustive. Following the classification of informal institutions that I have proposed:

(a) *Scientific conventions*. These are rules enabling and facilitating communication between scientists in order to make their discourse *intelligible* and *amenable to criticism*. It is only possible that pieces of evidence can bear on theoretical constructs, if both the evidential reports and the theoretical constructs are formulated with the aid of linguistic means that make them intelligible by others. Insights from mystical sources that cannot be shared with other members of the scientific community cannot be regarded as evidence. Only documentation of experimental results of excavation findings in archaeological sites etc., which are expressed with conventional, *standardized* means that make them intelligible to others can be part of the scientific discourse. The same is the case with theoretical constructions: those that are expressed in incomprehensible terms and which remain to a great degree private thoughts are not entities that can be shared with others and, thus, do not belong to the game of science. New concepts are, of course, introduced in scientific discourse embedded in theoretical structures – think only of such fundamental concepts such as force or phlogiston. But such novel concepts in the episodes of conceptual change[21] that one is familiar with in history of science are parts of larger theoretical constructs, which themselves remain largely intelligible by other members of the scientific community despite containing some new elements.

(b) *Specific moral rules.* Any human endeavour can be successful only if general rules of morality are respected. But for the specific task of bringing together theoretical constructs with evidence, specific moral rules for this specific epistemic task are more important than others. I want to especially stress the importance of *tolerance* and *sincerity*.[22] Science is a cosmopolitan arena where individuals and organizations from very different backgrounds participate. Producing the necessary evidence often requires the collaboration of extremely divergent agents – without tolerance such a collaboration would be impossible. Besides, the evidential exercise is a specific critical exercise with the means of evidence, but it remains a critical exercise, nevertheless. Dealing with criticism requires a tolerance on behalf of those that are being criticized for something in which they have often invested great parts of their professional lives. Besides, sincerity is the virtue, which is specifically important for the scientific endeavour, because it increases the epistemic trust that scientists can have in the testimonies of their fellows.[23] Since experiments can sometimes not be replicated due to reasons of cost or other reasons and since also reports from any source of evidence are to be trusted in order to play the evidentiary role that they are to play, sincerity is a *conditio sine qua non* for a smooth continuation of scientific activities. If no scientific agent were sincere or fraud were prevalent, the specific epistemic activity of bearing evidence on theoretical constructs would be impossible.

c) *Scientific techniques.* The logico-positivistic tradition in philosophy of science has produced arguments in favour of conceptualizing the way that evidence can bear on theoretical constructions as an *inference*.[24] The hope was that evidence formulated in observational statements could give support to a theory or hypothesis exclusively by using logical tools in the form of inferences. The scientific method was largely restricted to the algorithmic application of evidence expressed in propositions to infer the evidential support of a specific theory. What is worse, the case

> was made that this method encapsulated scientific rationality, so that an extremely restrictive view of human reason was implicitly endorsed. Duhem (1906/1981) was the first to challenge the slogan: scientific method = evidence + logic arguing that judgment is part of the scientific method[25] (for a discussion of his views, see the Excursus) and Popper (1934/2003, 1963 and 1972) presented in successive moves a critical rationalism, which was also wider than the slogan of the logical positivists.

This does not mean that inferences are not or should not be used in assessing how evidence bears on a theoretical construction. But in many cases, the very structure of the representational entities that are to be assessed, like, for example, diagrams or pictorial representations, do not allow the application of logical inferences at all. Besides, two problems exhaustively discussed in the literature make it obvious that genuine decisions are required on how evidence bears on theory, the underdetermination thesis and the argument from inductive risk. I will very briefly discuss them focusing exclusively on the core of the arguments (more extensive discussions are included in the Excursus).

The main version of the underdetermination thesis states that *all* theoretical constructs, that is, theories, models hypotheses etc. are underdetermined by logic and *all* possible evidence. This leaves a gap between evidence on the one hand and the acceptance of the respective theoretical construct on the other hand, which is inevitably filled by a choice.[26] This "Duhem-Quine thesis" is the stronger version of the thesis of underdetermination, also sometimes referred to as *global* underdetermination (Kitcher, 2001, p. 30f.; Biddle, 2013, p. 125). This thesis is often discussed with respect to theory choice, that is, with respect to the choice between alternative theories in the form that for any theory *T* there is always an alternative theory *T'* which is empirically equivalent to *T*. As Norton (1993, p. 1) and Dorato (2004, p. 58) stress, this claim is probably a figment of armchair philosopher's imagination due to "an impoverished picture of the ways in which evidence bears on theory."[27] The *transient* underdetermination thesis states merely that some theoretical constructs are underdetermined by logic and the *currently available* evidence. In the context of theory choice, this more moderate thesis states that at some moments in the history of science, there are some interesting cases of empirically equivalent theories (Psillos, 1999, p. 167). Insofar, as Norton (2008, p. 40) stresses, whether underdetermination is indeed the problem must be scrutinized on a case-by-case basis.[28]

The currently more popular argument from inductive risk that is discussed in different variants[29] focusses on the concrete decision that is involved when evidence is collected and the decision must be made whether the evidence is sufficiently high to warrant the acceptance of the hypothesis. The crux of the matter is that the fit between a theoretical construct and observations, the empirical fit of theory to facts, requires judgments about what kind of data to collect, when to stop the collection of data, how reliable are the data, what error margins are adopted etc. In a sense, there is not only one gap, but many gaps to be filled.

From the viewpoint of the approach defended here, it should be clear that what I call scientific techniques are the rules that provide the guidance to all those questions. In many cases, following these rules will solve the different aspects of these problems beyond any considerable doubt – there is not going to be any dissent about them,[30] and, in fact, there is going to be a straightforward application of the norms that the members of the scientific community have learned to follow over time. In other cases, where there are novel characteristics of the problem situation, they will have to make decisions. These decisions will diverge, because the truth-finding enterprise is at times excessively hard. And, of course, all these decisions will be fallible, dissensus will prevail and a critical discussion exactly about which data are relevant, when to stop collecting the data and whether the evidence warrants the theory will take place.

This discussion will be a value-laden discussion through and through. The decisions to accept a hypothesis, reject it or suspend judgment will be, however, reasonable decisions. Such decisions are *not arbitrary* decisions, but due to the tight institutional context of the enterprise, that is, the sum of all the rest of informal institutions that are in place, not every kind of reasons will count as scientific arguments in favour of accepting or rejecting a hypothesis. It is important to stress that even if some institutional rules are questioned some of the time, it is never the case that *all* informal institutions of science are questioned *at the same time*. If this were the case, then the participants would stop playing the game of science and start to engage themselves in other kinds of activities. The upshot of this is, that even if no algorithms are applied, the decisions made are *reasonable* decisions, that is, decisions encapsulating scientific rationality. It is simply untenable to equate calculation with rationality, and it is an impossible demand to require that judgments about which hypotheses should be accepted should be the outcome of calculations.

Finally, it is important to stress that the scientific techniques used to bear evidence on theoretical constructs evolve over time. New scientific techniques are invented, elaborated, and transferred often to areas where no one could originally preview that they could be used. The most obvious and celebrated case of a new scientific technique is, of course, experimentation. In contrast to medieval Aristotelism, experimentation was introduced as a novel way to gain knowledge during the Scientific Revolution. In the former framework, experimental intervention was regarded as a distortion of the natural course of events and seemed unsuitable to be used to gain access to the mysteries of nature. As Carrier (2013, p. 2554) emphatically remarks: "Johann Wolfgang von Goethe's criticism of Newtonian optics proceeded along the same lines. Goethe reproached Newton's experiments for forcing nature to respond in contrived ways so that scientists are misled in their interpretation. Experimentation is tantamount to torturing nature with the result that she gives deceptive answers. In order to obtain significant and trustworthy evidence about nature, she must be allowed to speak unconstrained." It took several hundred years to use experimentation in the social sciences. This is therefore quite spectacular, because for many decades during the twentieth century a standard argument for the alleged methodological dualism between the natural and the social (and human) sciences was the absence of the possibility of experimentation in the latter. Although psychological data were first obtained in a psychological experimental lab by Wilhelm Wundt in Leipzig in 1879, it was only at the end of the twentieth century that economic experiments were conducted in economic labs to become now along with field experiments and many other scientific techniques the standard repertoire of economists to bring evidence bear on models (Guala, 2009; Holt, 2019). In the biomedical sciences, randomized experiments are now considered the 'gold standard', especially where knowledge of the underlying mechanisms is very limited about the phenomena of interest,[31] but the invention of a controlled study where test and control groups are randomized so that the only systematic difference between them concerns the treatment that the members of the test group receive, has also been a novel, domain-specific scientific technique aiming at the warranting evidence.[32]

4.5 TRUTH AND SCIENTIFIC OBJECTIVITY

Scientific agents creatively construct representational entities that possess epistemic properties such as truth,[33] empirical accuracy, predictive power, simplicity etc. These properties are descriptive properties, which can be described with descriptive statements. I will start the discussion of the epistemic properties by focusing on truth, extending the argumentation that I have followed in Section 4.4. Recall that truth as correspondence to the facts is sometimes easy to find out in everyday life, when, for example, 'The cat is on the mat' is made true by the fact that the cat is on the mat. Since what science is after are vastly more complex states of affairs, finding out the truth is in most cases impossible. History of science is full of episodes where what was taken at a time to be accepted as certain knowledge turned out to be false in the light of new theoretical constructs. Scientists have become much humbler with respect to the aspirations that they could provide certain knowledge. This pervasive fallibilism of all our knowledge presses us to admit that these representational entities should be treated as hypotheses, amenable to criticism, conceptual and empirical. They can, however, be revised so that they can become, if not true, then closer to truth, truth-like as it were[34] or true enough.[35]

There is always going to be a gap between our representations and the facts, but how does the situation look like with respect to other descriptive properties of these representations? Empirical accuracy, for example, which can be further specified as *qualitative* accuracy (that is accuracy with regard to the existence of certain observable phenomena such as retrogressive planetary motion or cloud chamber tracks) or *quantitative* accuracy (that is accuracy with regard to observations, measurements or numerical outcomes of experiments) might be possible.[36] Internal consistency of a theoretical construct, that is the absence of contradictions, might also be possible. The fruitfulness of a theory, that is the prediction of novel phenomena might also be possible to state. And so on.

These epistemic properties other than truth are easier to describe, and they might be indicators of truth or markers of truth; this is not a priori given but must itself be found out. To put it differently, these descriptions of the epistemic properties of the representational entities are resources that scientists avail of in order to find out truth. In conditions of pervasive uncertainty where this novel problem must be solved, scientists make choices whether empirical accuracy, consistency, fruitfulness, etc. are such that the judgment can be warranted that a theoretical construct is close to truth or not.

When different theoretical constructs are assessed with respect to whether x is more accurate than y and y more accurate than z, then *comparative judgments* are made, which are *value judgments* comparing a property along some scale. It is surely confusing to speak in situations like these about the theoretical virtues *of* these entities or *of their* epistemic values; it should be clear that theoretical constructs of science are not bearers of virtues or values,[37] only scientists or other users of these constructs assign value to these properties, or to be more precise *evaluate positively or negatively these properties*. A lot of confusion can be avoided, if this simple point is acknowledged. It should be evident that different scientists evaluate these descriptive properties differently and this is what the scientific discourse is all about: it is an institutionally constrained dialogue about the different properties of the diverse representational entities, and about which one should be accepted among the ones on offer.[38] Part of the scientific discourse will be concerned about clarifications, that is, definitional issues, when the terms used to describe the representational properties are *vague*. Simplicity, for example, can be further specified as semantic or syntactic simplicity.[39]

It is in principle possible to make evaluative judgments with respect to a great variety of properties comparing the *relevance* of different properties of one representational entity towards enabling us to eliminate errors and come closer to the truth of this entity *or* the *relevance* of one property across different representational entities or finally, what is the hardest case, the *relevance* of different properties across different representational entities enabling us to state which one is closer to representing the facts. I will show in a more detailed manner how this can be done in Chapter 5 for the case of two main scientific activities, explanation and interpretation, exemplifying my argument with the help of case studies. For now, it is important to stress that such evaluative judgments can be made and are being made with respect to a great variety of properties, but *all* these properties are epistemic properties. The key point is that there are many properties that representational entities do not bear: a model cannot be emancipatory, sexist or racist. People can *use* it for emancipatory, sexist or racist purposes, but in itself a representational entity cannot have these properties. It is as if one would seriously claim that a stone *x* has the property of being empirically accurate.

But don't other values, normative resources at the highest, most abstract level, influence, very often crucially, which evaluative judgements with respect to the different epistemic properties of the representational entities will be made? Economic values, political values, feminist values, Marxist values etc.? They do. But it is a serious misinterpretation of the actual scientific practice to conceptualize the situation as if a scientist alone judges the evidential merit of a hypothesis and decides to accept it or reject it.[40] This decision is his or her *personal* decision, but not a *scientific* decision. It becomes a scientific one, only when it has passed through the tight institutional context of the scientific enterprise – *only then warrant is certified and objective knowledge is established.* A single scientist cannot be objective, only the rules of the scientific process and its outcomes can be objective.

The traditional ideal of scientific objectivity was strongly shaped by Francis Bacon and centred around the individual scientist without including the dynamics of interaction in a scientific community. Relinquishing one's prejudices, including alternative perspectives, and considering potential counterexamples were requirements on individuals conducting enquiries by utilizing one's senses. Error elimination was the aim of such an individual-centred account of objectivity. Baconian science was conceived as a social undertaking, to be sure, where division of labour was acknowledged: various features of a phenomenon can be illuminated by a number of researchers (Bacon, 1620/2000, Bk. I, § CXIII). But no particular significance was attributed to the *interaction* among scientists (Carrier, 2013, p. 2549). Rather the *individual application* of the Baconian rules of reasoning would suffice for attaining objective outcomes.

Objectivity is a normative, robust notion characterizing the process of assessment of the epistemic properties of the representational entities produced by scientists. According to my account, scientific objectivity should be conceived procedurally. The process is *more* or *less* objective depending on the extent to which the rules of the scientific method have been followed. These rules being conventional, moral and epistemic, scientific objectivity encompasses conventional, moral and epistemic dimensions.[41] Objectivity comes therefore in *degrees*, so that we can say 'process *a* is more objective than process *b*' depending on the extent to which the scientific conventions, moral rules and scientific techniques of a specific discipline have been followed.[42] So, when an authoritarian regime because of the extensive use of violence succeeds in substantially corrupting the range and intensity of the application of the scientific method, the respective process will be far less objective than otherwise.[43] Think of the case of Nazi science or Soviet science – these are extreme cases, where objectivity is minimal, if existent at all.[44]

A process exemplifying scientific objectivity to a high degree simultaneously reflects, thus, a great intelligibility of data and arguments, a nearly complete avoidance of fraud (Bright, 2017) and a careful application of the scientific techniques of a domain. The scientific community establishes the intersubjective validity of its products in virtue of following an objective procedure.[45] No procedure can be absolutely objective – a view from nowhere does not exist.[46] No procedure can include all participants – a view from everywhere does not exist either. The objective view of science is the view of no-one in particular (Fine, 1998).[47]

The intersubjective validity of propositions and other representational entities is a validity in terms of empirical fit, consistency, and fruitfulness, which has been attributed by scientists following the scientific method. We accept them because they have been through the critical scrutiny of the scientific community in a reasonable process, that is, a process exemplifying rationality. Errors of all kinds have been avoided to a greater or lesser degree thanks to the application of the scientific method, so that *impartial* outcomes have be offered.[48]

It is this impartiality that is most valued in scientific products. And it is important to understand that objectivity of the representations produced in science is a more complex, robust requirement, which goes beyond truth since it also includes the *appropriate use of skills* in experiments, measurements, etc., which cannot be truth apt, because they are not propositional. Besides, as I have stated earlier, there are representational entities like some types of models that are false, but still adequate for other purposes, like local predictions and the like. There can be warranted acceptability for such models on impartial grounds so that scientific objectivity can be attained to a higher or lower degree also in those cases.

This view of procedural objectivity that I propose allows for different degrees, is broader than truth and is useful for all representational entities produced by scientific agents. It is tied to the application of the scientific method: a process marked out as objective also allows the products to be characterized as objective. It is a more robust notion of objectivity than the one of logical positivism which conceptualized scientific method as the interplay of inference (deductive or inductive) and evidence (in the form of observation statements), stressing that inference has a shared logical structure, so that all scientists will draw the same conclusions from the same empirical data, and a widespread consensus will prevail.[49] My conception of the scientific method acknowledges the role of *reasonable choices* during the scientific process, the *diversity of domain specific techniques* and the specific moral rules of *tolerance* and *sincerity*.

The rules of the scientific method when learnt by the novices to a scientific community and anchored in their cognitive system will be applied quasi-automatically in all situations that are classified as "old problems" (according to the problem-solving framework presented in Chapter 3). With respect to a wide array of issues, a *consensus* in the form of shared mental models will be formed. The rules will evolve in a trial-and-error fashion once new conventions will be introduced and new domain scientific techniques will be invented. This view avoids the Scylla of absolute universalism and the Charybdis of relativism: these are not the only two options available as the logical positivists and the relativists contended. The scientific method is not timeless and ahistorical nor does it consist of random rules which are all equally good – the participants in the game of science share the rules of the scientific method, because they have learnt that they are the most effective available to describe the epistemic properties of representational entities. They can all be conceptualized as hypothetical imperatives: *if* you want to collect experimental data on the field, *then* follow rules *x* and *y*; *if* you want to assess how evidence in the form of empirical data provides support for a theoretical construct, *then* use the statistical rules k and m, etc. Rules of this type are neither strictly universal nor strictly local – these are general, often domain-specific scientific techniques serving the universal purpose of bringing evidence in contact with theory.[50] And they are fallible.[51] These general, fallible rules are to be understood as the rules of the game for successfully finding out the truth and the other epistemic properties of the diverse scientific representations (Albert, 1987; Sankey, 2021, p. 13ff.).[52]

When the rules will not prescribe an appropriate way to accomplish this task or when they will be in conflict, then individual choices will be made by individual scientists. These choices will be situated, however, in a tight institutional context and will not be arbitrary. And they will be criticized by other members of the scientific community, criticism remaining the main filter of arriving at intersubjective valid outcomes. Even when no agreement will prevail with respect to which values should have priority, a perfectly *rational* discourse is possible: this concerns the evaluation of the different rules with respect to different values, that is *a multidimensional evaluation* within the premises of a comparative approach. How this is possible will be shown in Chapter 5 for such entities as explanations and interpretations.

Concluding, the robust scientific objectivity proposed here is procedural, and it allows for degrees. During the epistemic problem-solving activities that scientists engage in, epistemic products are constantly spinning out of the process as the process itself evolves. The application of the scientific method bestows upon these products, which are mostly complex representational entities, intersubjective validity. These entities having passed through the tight institutional context of science enjoy scientific warrant to a different degree. All these diverse, warranted in different degrees, fallible representations constitute the stock of objective knowledge handed down from generation to generation. The process is not *neutral*, but the outcomes are *impartial* to a higher or lesser degree. They reflect the viewpoint of no one in particular. They are the most reliable representations of the natural and social world that we have because the process that has created and established them is a reliable one.

A Note on the Context of Application

I would like to close this chapter by briefly tying the main arguments of Sections 4.4 and 4.5 to the analytic distinction between context of discovery, context of justification and context of application. Epistemic problems are conceived and formulated in the *context of discovery*, and it should be clear that according to the conception of science defended here, all possible circumstances and all possible values exert their influence. Science is a social arena embedded in broader social structures, national and international, so that the scientific agents, individuals and organizations, will regard as *significant* a vastly diverse array of problems and will pursue their epistemic activities motivated by disparate motives.[53] It is impossible to construct significance graphs which could give orientation about the problems to be dealt with by scientific agents,[54] because the *evaluative structures* constantly change and there is no stability about what is to be regarded as significant – there is no external point of view or value system that would define significance somehow objectively, so that science will keep evolving spontaneously without intentional guidance by any external force.

The scientific products that have gone through the *context of justification* are warranted to a different degree. The warrant is the outcome of the critical discourse of the scientific community on the different ways that evidence bears on theoretical constructs. They are accepted provisionally as the fallible outcomes of an ongoing process, because they have passed evidential tests appropriate to the respective discipline. In innumerable cases, there is no possibility of application of a warranted finding for some practical purpose: finding out whether the universe expands, when dinosaurs went extinct from earth or when the technique of incising silhouetted figures in black-figured pottery was first used in Ancient pottery satisfy our curiosity about aspects of the natural and social world but cannot be applied to intervene in any meaningful sense. These findings can be used, of course, in other parts of science,[55] but they cannot be used for intervening in our natural and social environment.

In equally innumerable cases, warranted findings can be applied for practical purposes, and I will briefly focus in this note on the *context of application*. Here an obvious distinction should be kept in mind, which is systematically suppressed by all those who overdramatize the issue of application of scientific knowledge for practical purposes. This is the distinction between those cases where scientific warrant is beyond any reasonable doubt and those cases where epistemic products are in the process of being tested with data and an intensive dissensus about the evidence and their reliability prevails.

In the first case, when causal relationships are established, a technological transformation for practical purposes into the form of hypothetical imperatives is possible. This is the Weber-Popper-Albert proposal: factual statements can be turned with the appropriate logical transformation to technological systems, which indicate the appropriate means of intervention in the natural and social world.[56] If a scientific finding establishes a relationship between two phenomena of the type "every time that A occurs, then B occurs," then, if for practical purposes we want B to emerge, we have only to create the conditions for the appearance of A (Gemtos, 2016, ch.5). Take the case of the quantity theory of money, according to which the general price level of goods and services is proportional to the money supply in an economy – assuming the level of real output is constant and the velocity of money is constant. Suppose that we want to use the quantity theory of money for practical purposes. The statement: "Every time the money supply increases in an economy, and as long as GDP and the velocity of money remain unchanged, inflation increases proportionally" can be transformed into the equivalent statement: "If one wants to reduce inflation, then one has only to reduce money supply, as long as GDP and velocity of money remain constant." The actual decision is, of course, a political decision where a great array of values will play a role, but scientific agents need not make them qua scientific agents. No scientist is forced to make any political decision. As Carrier (2022, p. 17) emphasizes: "recommendations could be part of alternative conditionalized policy packages, each characterized by different goals."[57]

The stock of knowledge can in principle be used by any interested agent, individual or organization, to further whatever aims this agent tries to accomplish in any area of life, politics, markets, religion, arts or sports, exactly because of its universal characteristics.[58]

Since science is constantly evolving, epistemic products are constantly being created and their properties are debated in the scientific community. For a great array of them, there are diverse difficulties in collecting evidence, in testing them and in evaluating their epistemic characteristics. This is very often cutting-edge research or research where the scientific community suspends judgment because the evidence is insufficient or yet impossible to bear on theory. In all these cases, there is not only *uncertainty* which prevails – this is the case for literally all scientific knowledge – but *ignorance*: one simply does not know what the case is or what kind of causal relationships obtain. In all those cases, the demand from other actors in a society, political, economic, religious, etc. for scientific knowledge cannot be satisfied. But there is nothing peculiar or strange in such situations, since science is not a vending machine delivering epistemic products when one throws a coin. These demands will either not be satisfied at all, satisfied at a later point in time or satisfied by other systems of public beliefs available in a society. These cases of utter ignorance are not simple 'gaps' that can be filled by any kind of values to satisfy the respective demands. In the cases that some scientific agents pretend to do so, they violate sincerity, a constitutive part of the scientific method as I have argued earlier, and therefore fraud is involved.

Closing, a final remark to state the obvious: *doing* science is an activity different from science *communication*, science *education* and scientific *advice*. What is the appropriate place of these activities in a society organized as a polity is an issue that I will address briefly in later chapters. Let us now turn our attention to two central scientific activities, explanation and interpretation.

5 Core Scientific Activities

Explanation and Interpretation

Explanation and interpretation are two core theoretical scientific activities, and in this chapter, I would like to suggest a way of conceptualising them and of normatively appraising them within the premises of a comparative approach. There are many distinct activities taking place in the scientific process, but explanation is widely considered as the core epistemic activity in which representation and inferential reasoning are merged in a complex way in order to enlighten natural, biological and social phenomena. However, there is another core theoretical activity directed towards other goals that is equally complex: interpretation. This is the activity that deals with meaningful material. The importance of interpretation is standardly stressed in the humanities, but there is no reason to assume that this epistemic activity should be normatively appraised according to different standards than explanation. It is indeed my claim that there is no dichotomy of science and humanities at the methodological level and that the same normative approach can be successfully applied to both areas, notwithstanding a series of important differences. I will try to substantiate this claim in this chapter.

5.1 A DICHOTOMY BETWEEN SCIENCE AND HUMANITIES?

There are many facets of the complex problematic regarding the relationship between science and humanities and the debates between protagonists of these two domains of intellectual activity have been passionate both in the distant[1] and the recent past.[2] Partly due to the intensity of the debates and partly due to historical reasons, analytic philosophy of science has practically turned its back to the huge domain of the humanities. Philosophers of science standardly avoid dirtying their hands with the problems that scholars working in the humanities are confronted with. There is indeed an implicit consensus of problems accepted as legitimate objects of consideration by philosophy of science, a consensus implicitly demarcating science from the humanities. The core epistemological issue which underlies the most convincing attempts at dichotomising science and humanities is the question of *how to deal with meaningful material*. Human actions are meaningful and texts and other byproducts of human action constitute meaningful material. How are they to be treated? Analytic philosophy of science has – with very few exceptions – systematically avoided the treatment of the concrete problems that emerge when dealing with meaningful material, so that literally no attention is paid to a great range of disciplines that deal with the interpretation of such material.[3]

Portraying scientific activity as primarily explanatory activity builds on a long tradition in philosophy of science (Hempel, 1942/1965; Popper, 1934/2003). A series of very sophisticated theories of scientific explanation have been offered in the last decades (e.g., Friedman, 1974; Kitcher, 1981, 1989; Salmon, 1984; Machamer, Darden, and Craver, 2000; Woodward, 2003; Craver, 2007; Strevens, 2008), and the initial focus on explanatory arguments has been abandoned in favour of a focus on explanatory practices (e.g., Love, 2015; Woody, 2015). Philosophers analysing explanatory activities have recently been inclined to adopt pluralist positions, and although the field of the philosophy of explanation is constantly moving forward (e.g., Lange, 2016; Khalifa, 2017; Plutynski, 2018; Skow, 2018; Rice, 2021), the implicit consensus has remained stable: explanation is the 'royal' scientific activity in which practices of scientific representation and accounts of scientific inference are skilfully merged with considerations regarding the appropriate scope of application to shed light upon natural, biological and social phenomena.

However, it is an enormous bias to portray all and every theoretical scientific endeavour as explanatory activity aiming at answering 'why?' questions. If the activities of historians, archaeologists, classicists, philologists, etc., are not to be denied the honorific title 'scientific',[4] then another central epistemic goal should be recognised: the ascertainment of facts that answer 'what is the case?' or 'what was the case?' – questions. The answers to questions of this type allege the existence of facts: they are most often singular descriptive sentences about event-tokens and are temporally and spatially determined. The activity that leads to the production of such answers is an interpretative activity.

The purpose of this chapter is on the one hand to show the use of scientific method *in concreto* and on the other that it is applicable both to the sciences and the humanities. In order to keep the task as manageable as possible, I will focus on what is regarded as the 'royal' epistemic activity in the sciences, that is, explanation, and I will juxtapose it with what is regarded as the 'royal' epistemic activity in the humanities, that is, interpretation. This is not to negate that there are other important activities undertaken in the sciences, for example, prediction or control or that there are other important activities undertaken in the humanities, for example, the active promotion of ethical or civic sensibilities. However, *explanation* and *interpretation* are recognised as the core theoretical activities in the sciences and in the humanities, respectively, so that they can serve as focal points structuring my discussion.

I will articulate and defend two interconnected claims: firstly, that the main activity of theoretical science is *explanation*, that the main activity of the humanities is *interpretation*, and that they can be both conceptualised as *epistemic problem-solving activities*; secondly, that both explanations and interpretations can be normatively appraised within the premises of a comparative approach using the *same standards* involving intersubjective intelligibility, testability with the use of evidence and rational argumentation, so that there is not any *in principle* difference between science and humanities – *a dichotomy* of science and humanities at the methodological level *is*, thus, *untenable*.

In Section 5.2, I will deal with explanatory problems and explanatory activity showing how the complex practice of explanation should be conceptualised. In Section 5.3, I will reconstruct interpretative activity as epistemic problem-solving activity showing how exactly interpretation constitutes a distinct epistemological issue due to the complex problematic of meaning. Section 5.4 will show how the normative appraisal of explanations and interpretations can take place satisfying the standards of intersubjective intelligibility, testability with the use of evidence and rational argumentation and therefore how they may exemplify scientific rationality. The chapter will close with a short conclusion in Section 5.5.

5.2 EXPLANATORY PROBLEMS AND EXPLANATORY ACTIVITY

A fruitful way to analyse the explanatory enterprise in science is to shed light on the social *process* of explanation which unfolds in historical time instead of focusing on the *outcomes* of this process and their characteristics. Highlighting the complex process of explanation rather than posing the issue as if explanation were a static situation can be done by employing the notion of an explanatory game.[5] Scientific agents provide solutions to explanatory problems by following four kinds of rules: (a) *constitutive rules*, the basic set of rules that constitute an explanatory game as a game, (b) *rules of representation*, which can be linguistic, diagrammatic, pictorial, acoustic, etc., (c) *rules of inference*, which comprise the inferential strategies used by the agents and (d) *rules of scope*, which define the scope of phenomena to which the explanatory game should be applied. By following these rules, explanations are constantly produced in a process of trial and error aiming at the solution of the explanatory problems that the scientific agents are facing.

According to this view, an analysis of the fictitious entity called 'the explanatory relation' is a dead end. If one is prepared to abandon the stance that explanation is an outcome in favour of a stance that explanation is an activity, then a whole range of new issues emerges.

I will focus here only on one issue, that is, how the representation of an explanatory problem takes place and how this is connected with interpretation.

An agent is confronted with a problem when (s)he wants something, but (s)he does not know what series of actions (s)he can perform to get it (Newell and Simon, 1972, p. 72). Problem-solving can be treated "as a process of searching through a *state space*. A problem is defined by an *initial state*, one or more *goal states* to be reached, a set of *operators* that can transform one state into another, and *constraints* that an acceptable solution must meet" (Holland et al., 1986, p. 10). Explanatory activity can be analysed as a special kind of problem-solving, in which the goal state is a target to be explained. There are a variety of ways that an explaining agent may obtain the goal state, depending on the diverse forms that the explananda can take. Constructing deductive arguments is one such way, but very often this is impossible. In medicine, for example, visual hypotheses are often employed about the shape and location of a tumour in order to explain observations that are represented not only by words, but also by the use of touch, smell and sight (Thagard, 2012, p. 37).

Representation is, thus, an integral part of explanatory problem-solving.[6] The most common form of representation is *linguistic representation*. The explanatory problem is very often represented by using rules that organise symbols, usually letters which are identifiable by their shapes, and fix their meaning. The fact that the visible forms of written sentences are arbitrary with respect to their meaning (beyond the specific rules that fix it) is the main source of their convenience and almost universal employment in common sense explanations and in many scientific explanations. *Visual representations* are further forms of representation and can be mainly of two kinds: *diagrams* which avail of an articulate syntax, but also of a semantics that assigns referents to, in principle, unambiguously identifiable characters (Perini, 2005a) and *pictorial representations* which do not avail of an articulate syntax, since it is impossible to tell exactly what character a mark instantiates (Perini, 2005b, p. 916).

There are also *acoustic* representations (Palmieri, 2012), as well as representations of *taste* and of *odours*.

Giere correctly argued against a dyadic notion of representation (2006, p. 60): "I suggest shifting the focus to scientific practice, which implies that we should begin with the practice of representing. If we think of representing as a relationship, it should be a relationship with more than two components. One component should be the agents, the scientists who do the representing. Because scientists are intentional agents with goals and purposes, I propose explicitly to provide a space for purposes in my understanding of representational practices in science. So we are looking at a relationship with roughly the following form: S uses X to represent W for purposes P. Here S can be an individual scientist, a scientific group, or a larger scientific community. W is an aspect of the real world. So, more informally, the relationship to be investigated has the form: Scientists use X to represent some aspect of the world for specific purposes."

The specific purpose in our case is the solution of the explanatory problem at hand. There is, as we have seen earlier, a great variety of possible representations of an explanatory problem (Frigg and Hunter, 2010). I would like to suggest that representation is an essentially triadic relationship including three relata: the representation-bearer, the representational object and the interpretation in a cognitive agent. Representation is constituted by these three relata and the relations obtaining among them.

There are different entities that can count as representation-bearers, like linguistic expressions, diagrams, computer monitor displays, photographs, etc. "A representation-bearer is a physical object which is about (directed at, stands for, refers to, points to) an object or state of affairs other than itself" (Files, 1996, p. 400). This is the most fundamental feature of representation and what distinguishes representational entities from non-representational entities. The representational object is the object or state of affairs that a representation-bearer is about. It is *what* the representational bearer is directed at, stands for, refers to, points to. The metal plates along with rods arranged helically around a stand represent DNA by virtue of sharing the same shape and structure. The so-called Newlyn-Phillips Machine, a hydraulic machine built in 1949 by two economists, Walter Newlyn and Bill Phillips, represent the macro-economy (Morgan, 2012, p. 176ff.). There is, however, nothing automatic yielding the representation when representation-bearers are directed at, stand for, refer to or point to a representational object. A cognitive agent is required to establish the connection between the two by denoting, in the case of linguistic expressions, and by selective resemblance[7] in the case of virtual representations, a combination of both or by entirely other ways (van Fraassen, 2008, Part I; Daston and Galison, 2007, ch. 7). Interpretation by a cognitive agent is, thus, a constitutive part of a triadic relationship of representation as is shown in Figure 3.

The cognitive agent can be an individual or a group of individuals (a research group or a scientific community). Interpretation is thus a constitutive part of the process of representation taking place during explanatory problem-solving. *There is no explanation without representation and no representation without interpretation.*

In order to provide a concrete example, consider the explanatory problem-solving activities regarding the functioning of the heart and the circulation of the blood and how they have evolved in the explanatory game that unfolded from Galen to Harvey. The rules of representation that aimed at representing heart and blood circulation were initially limited to the use of natural language by Galen himself, without the use of dissection (in Galen's time dissection was only permissible to animals). The reappearance of human dissection at the

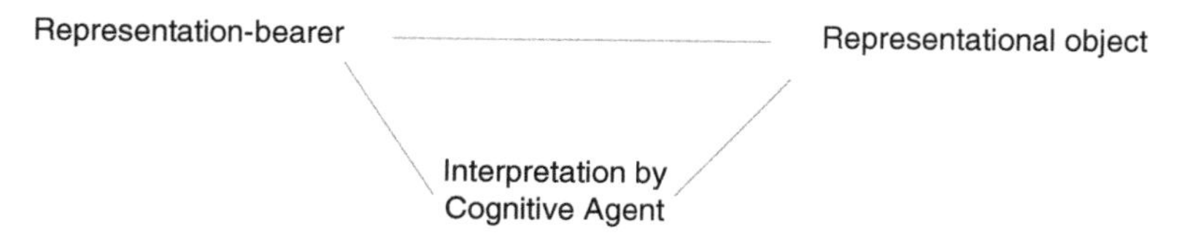

FIGURE 3 Representation as a triadic relationship

University of Bologna took place by Mondino around 1315 (Siraisi, 1981, 110ff.). Vesalius' publication of *De Humani Corporis Fabrica Libri Septem* (On the Fabric of the Human Body) in 1543 was a further milestone – he collaborated with the workshop of Titian to construct woodblock engravings of the drawings contained in the *Fabrica*. This in turn enabled the representation of the heart by drawings and other illustrations which could be copied and multiplied with accuracy using printing technology. The crucial move in the explanatory game consisted in the novel explanatory problem-solving activity undertaken by William Harvey in 1628 in his *Exercitatio Anatomica de Motu Cordis et Sanguinis* (Harvey, 1628/1989).

The rules of representation kept evolving further: the use of microscope, first employed in the middle of the seventeenth century by Malpighi (1661/1929), made possible the accurate depiction of the anatomy of the heart. Other types of rules, like the rules of inference and rules of scope, have also evolved over the centuries, exemplifying the historicity of the scientific explanatory game.

Summing up, the rules of representation that explainers follow in their problem-solving activities comprise (a) rules determining which entities count as representation-bearers, (b) rules determining by virtue of what a representation-bearer is supposed to represent and (c) rules determining by virtue of what a representation-bearer is connected with the representational object. Apart from *rules of representation*, explainers also use *rules of inference* acting on representations; different types of representation allow for different kinds of inference. To take an example, Euclidean geometry shows that rigorous inferences can be drawn on strictly visual representations. And many types of formal inferences can be performed on linguistic representations, like mathematical expressions. Finally, explainers also follow *rules of scope* that delineate the scope of phenomena to which the explanatory endeavours should apply. Explanatory activities unfold, thus, in a collective problem-solving process during which participants in the explanatory game follow rules of representation, rules of inference and rules of scope. Explanations are permanently generated as an outcome of the interaction of the explainers combining and synthesising these rules in their activities. I will defer the discussion of the normative appraisal of explanations to Section 5.4 and will now turn to interpretative problems and interpretative activity.

5.3 INTERPRETATIVE PROBLEMS AND INTERPRETATIVE ACTIVITY

Analytic philosophy of science has systematically neglected to pay attention to a great range of problems that scholars working in the humanities are confronted with. The purpose of this section is to briefly reconstruct the main practice of the humanities and then, in Section 5.4, to normatively appraise its products, along the lines presented in the last chapter. We have seen earlier that interpretation is a constitutive part of representation and thus part of explanatory problem-solving activities. This is probably the reason why philosophers of explanation were inclined to underestimate the importance of interpretation for decades, reducing or subsuming it under a more inclusive explanatory enterprise aiming at the answering of 'why?' – questions. However, in many disciplines of the humanities the task is to provide satisfactory answers to 'what is the case?' – or 'what was the case?' – questions, which are not necessarily a step towards providing in the end satisfactory answers to 'why' questions.

To use the triadic relationship of representation invoked in Section 5.2, the scientific work of scholars in the humanities consists of a proper analysis of the relationship between the representation-bearer and the representational object by engaging themselves in interpretative activities. These activities are often much more laborious than in the natural, biological and social sciences, because of the complexity of this relationship. The main source of complexity is that the material dealt with is *meaningful*. In other words, in the humanities the interpretative task becomes so dominant that the whole scientific effort is essentially focused on accurately establishing the link between a representation-bearer and the representational object. The central epistemological goal is a distinct one of answering 'what is the case?'-questions and, thus, of solving exclusively *interpretative problems*.

In the year 1901, a vessel carrying professional sponge-divers was driven by a storm to a remote island in the Aegean Sea called Antikythera. While fishing in the area they discovered the remains of an ancient shipwreck carrying a rich cargo of sculptures, glassware, pottery as well as other artifacts. Among the findings was one of the most curious objects ever found in an archaeological operation: a metallic construction hardly discernible among the surrounding rocks, consisting of a complex arrangement of bronze gears intricately intertwined and compactly encased in some kind of a container. A member of the crew took it to be a useless piece of rock, and it was only the young naval officer who prevented him from throwing it back into the sea. What was this device? Researchers over many decades have tried to shed light on the device's construction and purported operation. Freeth et al. (2008), in an article published in *Nature*, maintain that this so-called Antikythera Mechanism must have been constructed during the second half of the second century BC in some major centre of scientific research of the time, possibly in Syracuse in Sicily.

The representation-bearer is this rather inconspicuous metallic device, the representational object is unknown, and the cognitive agent is the archaeological community. The *interpretative problem* consists of (a) successfully reconstructing the representation-bearer; (b) successfully specifying the representational object, that is, what the device was supposed to represent at all; (c) inventing hypotheses on how exactly (a) and (b) are connected. The main epistemological goal is to solve this interpretative problem. This does not necessarily exclude the possibility of also trying to solve explanatory problems deemed interesting and important in a second step, but interpretative problem-solving is a largely autonomous activity here, precisely because of its complexity.

The first part of the interpretative activity aims at successfully reconstructing the *representation-bearer*, unfolded by virtue of identifying the nearly thirty gears in the remaining fragments and of estimating the number of missing gears and how they all comprised the Antikythera Mechanism. This involved the use of two non-destructive investigatory techniques: *Polynomial Texture Mapping* (Malzbender and Gelb, 2006) to enhance surface details of the fragments, and *Microfocus X-ray Computed Tomography* to examine the interiors of the fragments at high resolution. Delineating the *representational object* makes up the second part of interpretative activity. Freeth and Jones (2012, § 2.3.2) maintain that the Mechanism combined two main functions "as an analogue computer, permitting quantitative read-off of the longitudinal positions and motions of the heavenly bodies, and as an educational wonder-working device, portraying the cosmos and its constituent parts in their hierarchical structure and intricate movements. The entire complex of dial and pointers on the Mechanism's front thus could by metonymy be itself called the 'cosmos', and we are convinced that this is what the word *kosmos* in line 25 referred to. Though *kosmos* had a range of possible meanings outside of scientific contexts, in Hellenistic astronomy it always meant either the aggregate of the heavens and the Earth or the heavens as distinct from the Earth."

The third part of interpretative activity concerning the ways that the representation bearer is connected with the representational object is even more complex. In a detailed study of Platonic astronomy, Kalligas (2016) has proposed the hypothesis that the construction of mechanical devices (σφαιροποιία) was a constitutive part of ancient astronomy, already referred to in the work of Theon of Smyrna, *On Mathematical Issues Useful for the Reading of Plato,* of the second century AD. The Antikythera Mechanism is, thus, according to this hypothesis, such a device (the representation bearer) representing the motions of the celestial sphere (representational object) in the tradition of a quite sophisticated astronomy traced back to Plato (connecting the representation bearer to the representational object). Ongoing research on inscriptions specifying complex planetary periods constantly forces new thinking of the mechanisation of the Cosmos in the Antikythera Mechanism, as in a new article published in *Nature Scientific Reports* where photographs and illustrations are readily accessible (www.nature.com/articles/s41598-021-84310-w).

There are interpretative problems that are even more complex. These are cases in which the representation bearer itself consists of material where the problematic of meaning appears accentuated. I will briefly deal with the problem of *text interpretation*, precisely because it is a hard case and because it has provoked solutions in the last decades that seem to violate precepts of scientific rationality. Here the representation-bearer is a text, the representational object can be a fact, a state of affairs, a fictional entity or an aspect of a fictional world and the activity establishing the link between the text and the representational object constitutes the text interpretation. A text consists of a more or less structured series of linguistic expressions. Linguistic expressions are bestowed with meaning by their author while (s)he construes them against the background of his/her goals, his/her beliefs, and his/her other mental states in the process of interacting with his/her natural and social environments; such a construal of a *nexus of meaning*[8] is a complex process involving the conscious and unconscious use of symbols. The scientific problem of interpretation consists in answering the specific 'what is the case?'-question, that is, in describing the process of construal of the nexus of meaning by its author, in depicting the representational object at which the nexus of meaning of the text is directed or to which it refers and in analysing the connections between representation bearers and representational object.

It is important to stress that text interpretation is a distinct activity from highlighting the *significance* of a text. Text interpretation is essentially about *reconstructing the nexus of meaning of a text*; textual criticism is an activity drawing attention to the significance of a text with respect to specific goals and purposes. Authorial self-repudiation exemplifies this distinction well: an author's later disapproval of his/her own work can be in sharp contradiction to his/her original appreciation – this shows that though *the meaning of the text has remained stable*, the significance of the work to the author has changed dramatically. Besides, text interpretation and *application* of a text are also distinct activities: applying a legal text in the field of jurisprudence aims at regulating specific social problems and applying a literary text aims at adopting a particular view of life on the part of the reader – these applications have nothing in common with an accurate reconstruction of the nexus of meaning of the respective texts.

In order to *re*construct the original nexus of meaning of a text one has to invent respective hypotheses. Such reconstructions can be carried out with the most diverse conceptual apparatuses, and they are hypothetical insofar as it is uncertain whether they are felicitous to meeting the epistemological goal, that is, to identifying the nexus of meaning of the text, or not. As in the analogous case of answering 'why?'-questions, there are no particular algorithms underlying the process of formulating interpretative hypotheses that can adequately reconstruct the nexus of meaning. The interpretative principles that have been proposed in the older discussions on "radical interpretation" like the *principle of charity* (Quine, 1960, p. 59; Davidson, 1984, p. 27) or the *principle of humanity* (Grandy, 1973) can be viewed as *presumptive rules* that can break down in the light of experience.[9] In other words, one does not need to commit to these principles being constitutive for the practice of interpretation. Their apparent indispensability is to be traced to the fact that they have been very often employed with success. They remain hypothetical, nevertheless, and they can always fail to stand up to experience.

It is important to stress that the interpretative hypotheses might not refer to directly observable entities, as this is often the case in the explanatory practices in the natural, biological and social sciences. The interpretative hypotheses very often consist of statements about the *intention* of the author, for example, which is surely nothing directly observable. However, one can deduce from such interpretative hypotheses, in conjunction with other statements, consequences that are more observable. Once the interpretative hypotheses are formulated and observable consequences are drawn from them by means of deduction, it is possible to empirically test them – the interpretative practice is, thus, to a great degree an evidential exercise (Gomperz, 1939, 59ff.). This evidence consists of details of rhyme, rhythm, along with other stylistic means as well as of what the author says about her work, biographical evidence, the broader intellectual and the narrower linguistic context of the text (Skinner, 1969, 1972, 1975; Føllesdal, 1979; Nehamas, 1981, 1987; Mantzavinos, 2005, 2014). As Rescher (1997, p. 201) correctly stresses: "The crucial point, then, is that any text has an envisioning historical and cultural *context* and that the context of a text is itself not simply textual – not something that can be played out solely and wholly in the textual domain. [...] The process of *deconstruction* – of interpretatively dissolving any and every text into a plurality of supposedly merit-equivalent constructions – can and should be offset by the process of *reconstruction* which calls for viewing texts within their larger contexts. After all, texts inevitably have a setting – historical, cultural, authorial – on which their actual meaning is critically dependent."

In order to take a concrete example, consider the problem of interpretation of the invisible hand in Adam Smith's *The Theory of Moral Sentiments* (1759/1976), but mainly also in his work *An Inquiry Into the Nature and Causes of the Wealth of Nations* (1776/1976, p. 477f.), a problem which has been salient in scientific and political discussions: "[Every individual] generally, indeed, neither intends to promote the public interest, nor knows how much he is promoting it. By preferring the support of domestic to that of foreign industry, he intends only his own security; and by directing that industry in such a manner as its produce may be of the greatest value, he intends only his own gain, and he is in this, as in many other cases, led by an invisible hand to promote an end which was no part of his intention. Nor is it always the worse for the society that it was no part of it. By pursuing his own interest, he frequently promotes that of the society more effectually than when he really intends to promote it. I have never known much good done by those affected to trade for the public good. It is an affectation, indeed, not very common among merchants, and very few words need be employed in dissuading them from it."

There are three main interpretations: (a) a standard interpretation aiming at the reconstruction of the nexus of meaning of this passage according to which a society of self-interested individuals constrained by criminal law and the law of property and contract is capable of an orderly disposition of its economic resources (Hahn, 1982, p. 1); (b) a non-standard interpretation which additionally highlights the use of "the invisible hand of Jupiter" by Adam Smith in a passage in his *History of Astronomy*, originally published in his *Essays on Philosophical Subjects* (1759/1980, p. 48f.) where the invisible hand seems to disturb the 'orderly course of things'. Macfie (1971, p. 596) argues that while the capricious role of the 'invisible hand of Jupiter' is quite different from that of the order-preserving 'invisible hand' of the *Wealth of Nations* (and of the *Moral Sentiments*), there is no inconsistency since it reflects the view of history typical of the Enlightenment. Jupiter's invisible hand representing the ignorant 'savage's' view and the invisible hand in the Wealth of Nations both "in fact describe Adam Smith's interpretation of how the natural order of 'providence' animates and directs the orderly development of these societies (as well, of course, as the physical universe). One is reminded of the first and third ages of Vico – the age of gods and the age of men. But similar interpretations of history were common in the Scottish eighteenth-century school" (ibid.) This interpretative hypothesis is constructed following the principle of charity, since it is assumed that all passages of the author, Adam Smith, are in general internally consistent and consistent with one another. This allows him to present a hypothesis concerning the invisible hand that provides a unitary, consistent interpretation and makes a number of different and specific features of the work all appear fitting; (c) an ironic interpretation of the invisible hand according to which Adam Smith did not particularly esteem the invisible hand and thought of it as an ironic though useful joke (Rothschild, 1994, 2002).

All three interpretations consist of statements about the intention of the author and are based on different sorts of evidence which help to *reconstruct the nexus of meaning* of the respective text.

The case of text interpretation is a pure case of an interpretative problem and an interpretative problem-solving activity. Such an activity will yield epistemic products, that is, interpretations which can be normatively appraised using the same standards employed in the sciences as will be shown in the next section.

5.4 EXPLANATION AND INTERPRETATION: NORMATIVE APPRAISAL

In the last two sections I have portrayed explanatory activities aiming at answering 'why?'-questions and interpretative activities aiming at answering 'what is the case?'-questions as distinct kinds of epistemic problem-solving. They are both trial-and-error procedures oriented towards the achievement of two distinct central epistemological goals. I have stressed that they are both hypothetical, precisely because it is uncertain whether the problem solutions generated will indeed *successfully* meet the epistemological goals. Whether this is the case and what constitutes success can be shown by undertaking a systematic normative appraisal to which this section is dedicated.

Not all explanations are equally good. There are better and worse explanations. This trivial point is standardly acknowledged by scientists in their daily work. Are all interpretations equally good? A scientifically trained mind would immediately answer negatively. However, a series of scholars in literary theory and other humanistic disciplines and a series of postmodern philosophers answer positively. According to deconstructionism, for example, any reader of a text establishes an interpretation of the text and one reader's interpretation is supposedly as good as another's. This view is very problematic as Nehamas (1981, p. 140f.) stresses: "Derrida argues that even the most obvious reading is the result of interpretation and can therefore be questioned, revised, or displaced. This is, I think, correct. Just as in scientific explanation there are no data immune to revision, so in literary criticism there are no readings impervious to question. But the fact about science does not show that apparently competing scientific theories are incommensurable and that therefore we cannot judge between them or that each such theory concerns its own distinct world. Similarly, the point about criticism does not show that different interpretations of a text are, even if apparently incompatible, equally acceptable or that a text has as many meanings as there are interpretations of it. Readings are neither arbitrary nor self-validating simply because they are all subject to revision. Newer readings are always guided by the strengths and weaknesses of those which already exist; and though this process may never stop, it is not for that very reason blind."

As we have seen in the last chapter, the epistemic properties of the entities produced in science can be described by descriptive statements. To find out the true explanation or the true interpretation might be very difficult on many occasions, but other epistemic properties could be markers or indicators of truth: empirical accuracy, consistency, coherence, simplicity, fruitfulness etc. Scholars then make choices whether these epistemic properties are such, that the judgment can be warranted that a theoretical construct like an explanation or interpretation is close to truth or not. When different explanations or interpretations are assessed with respect to whether *x* is more accurate than *y* and *y* more accurate than *z*, then comparative judgments are made, which are value judgments, *evaluations* comparing a property along some scale. Such judgments are made within a normative dimension in which the different rules guiding the explanatory or interpretative activity will be evaluated with respect to values which encapsulate the normative resources at the highest level.

A series of those values will be shared in the respective scientific community, so that a great part of the discussion will revolve around which kind of rules can attain which kind of values. Such statements can be perfectly unambiguous. As Laudan (1984, p. 31f.) correctly observes in this context: "Suppose that a scientist is confronted with a choice between specific versions of Aristotle's physics and Newton's physics. Suppose moreover that the scientist is committed to observational accuracy as a primary value. Even granting with Kuhn that 'accuracy' is usually not precisely defined, and even though different scientists may interpret accuracy in subtly different ways, I submit that it was incontestable by the late seventeenth century that Newton's theory was empirically more accurate than Aristotle's. Indeed, even Newton's most outspoken critics conceded that his theory was empirically more accurate that all its ancient predecessors. Similarly, if it comes to a choice between Kepler's laws and Newton's planetary astronomy and it does come to such a choice since the two are formally incompatible and if our primary standard is, say, scope or generality of application (another of the cognitive values cited by Kuhn), then our preference is once again dictated by our values. At best, Kepler's laws apply only to large planetary masses; Newton's theory applies to all masses whatsoever. Under such circumstances the rule 'prefer theories of greater generality' gives unequivocal advice."

Now, it is the case that due to the long apprenticeship and socialisation of scholars in their respective communities, they tend to share a series of values; this is the reason that *evaluations* of the different types of rules that they follow in their daily epistemic activities with respect to a series of values will be unambiguous. But there are going to be cases where a dissensus will prevail about which values should be weighed more. It is a common claim among literary theorists, for example, that interpretations can and should not be evaluated in reference to truth, but rather aesthetic values. This claim is both plausible and correct. The same set of interpretations can be evaluated with respect to different values, such as beauty or originality, and found to be fulfilling some of them while not fulfilling others. An interpretation can be, for example, original but at the same time false. However, the situation is not fundamentally different in the sciences. Some explanations might be novel, but false.

A pluralistic account is tenable for both the humanities and the sciences and is an account that acknowledges positively evaluations of diverse *epistemic* properties – like, empirical adequacy, simplicity,[10] accuracy, consistency, etc. which are all markers of truth. When truth is not possible to find out, especially at the beginning of the scientific inquiry of a subject matter, then the evaluation of other epistemic properties will be the only possible alternative. *Non-epistemic* properties – mainly of aesthetic nature, like symmetry, harmony or the metre of a text can also be considered. *Pluralism* (as a normative account, opposed to plurality as a positive description of a state of affairs) is fundamentally different from relativism, however, since it involves neither a resignation nor a renunciation of commitment, as relativism does (Chang, 2012, p. 261). Besides, the relativistic requirement to treat all alternatives as of equal merit is distinct from the requirement to propose and discuss multiple alternatives. Such an approach goes hand in hand with *fallibilism*, which acknowledges that all our activities, knowledge and principles are prone to error. Human beings are constantly making mistakes in all areas of cognition and praxis, but they are able to learn from their mistakes. Insofar, all our problem solutions and all our evaluations should be treated as hypothetical, but criticisable and, thus, revisable.

The plurality of rules guiding the problem-solving activities and the plurality of epistemic and non-epistemic properties call for a multidimensional evaluation.[11] Different sets of rules can be evaluated with respect to the degree of exemplifying different properties – this is the main credo of the proposed approach. *Appraisal is always a comparative matter* – we always form evaluative judgments by comparing different alternatives. Such comparisons are to be juxtaposed to evaluations undertaken within the premises of an ideal or even transcendental approach.[12] The notion of an 'ideal explanatory text', influentially proposed by Railton (1981, p. 240ff.) in the philosophical theory of explanation, is such a characteristic attempt to work out an ideal standard. The intuition behind the comparative approach, which is founded on the claim that the availability of an ideal approach is neither sufficient nor necessary for a comparative evaluation, can be best captured by an analogy: the fact that a person

Evaluations

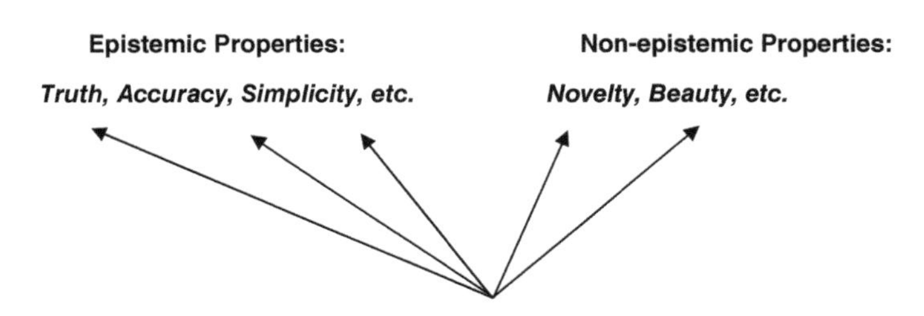

Rules of representation, rules of inference, rules of scope, interpretative rules etc.

Explanatory and Interpretative Problems

FIGURE 4 The comparative approach to evaluation

regards the *School of Athens* as the best fresco in the world does not reveal how (s)he would rank a Monet against a Rubens. In order to make a judgment that *x* is better than *y*, we do not need to maintain that some quite different alternative *z* is the best one. Figure 4 gives a diagrammatic representation of the comparative approach to evaluation.

A multidimensional evaluation is naturally possible. Decades of discussion after the heyday of logical positivism in the analytic philosophy of science have brought to light the impossibility of unequivocally making supposedly rational decisions about the validity of scientific knowledge by *applying specific algorithms*. Such decisions are naturally problematic also in the case of the evaluations of explanations and interpretations with respect to different epistemic and non-epistemic properties. Constructing algorithms and sometimes even calculi designed to replace controversies with calculations is a futile enterprise. Basing methodology merely on the interplay of logic and evidence is surely insufficient, since it unduly brackets the *role of imagination* in the cognitive praxis and probably most importantly the *role of choice*.

Creative decisions, that is, human decisions by real-world human beings, are permanently involved in all stages of the process of comparative evaluation of alternative explanatory and interpretative hypotheses. Dethroning Olympian rationality as an impossible ideal does not necessarily lead to irrationalism or relativism, however. The rationality of the proposed comparative approach does not lie in the proof or defence of a specific algorithm or justificatory procedure as the most adequate one to establish the goodness of an explanation or interpretation. The rationality lies instead in laying the ground for the criticism of the diverse explanatory and interpretative hypotheses.

I would like to endorse a *procedural conception of rationality*[13] here which does not prescribe the content of the decisions ultimately taken up by the participants of the discussion about the appropriateness of certain explanations and interpretations themselves. It highlights instead the individual and collective conditions that must prevail in order for these decisions to be informed and adequate. These decisions are themselves fallible. Institutionalising the possibility of criticism is the best means to facilitate the correction of errors when choices are involved. The anchoring of the *freedom of criticism* in the *institutional framework* where such problem-solving activities are conducted is the broader collective condition that must prevail and enables procedural rationality to manifest itself. Our fallible decisions are all that we have here as elsewhere and enabling a critical discussion is the prerequisite of making informed choices about which explanations and interpretations to provisionally adopt. And it is important to stress again that there is no algorithm relieving us from the necessity to make the choices. Controversies cannot be substituted by calculations – we still have to form judgments and decide which calculations to adopt.

In a nutshell, good science is the outcome of both sound reasoning and successful decisions, on the one hand, and of appropriate institutions allowing for the possibility of criticism to get around the cognitive and motivational limitations of scientists and to correct their error-prone problem-solving activities, on the other. The same applies to the humanities: good epistemic work here is also the outcome of both sound reasoning and successful decisions, on the one hand, and of appropriate institutions allowing for the possibility of criticism to get around the cognitive and motivational limitations of scholars in the humanities and to correct their error-prone problem-solving activities, on the other. Consequently, there *is not any in principle* methodological difference between science and humanities, if a comparative approach of normative appraisal is accepted and a view of procedural rationality is endorsed.

5.4.1 Normative Appraisal of Explanations

I would now like to show how such a comparative approach of normative appraisal is substantiated in the two cases of scientific problem-solving activities discussed in this chapter, explanation and interpretation. This will be done with the help of the examples introduced to portray explanatory and interpretative problem-solving activities in Sections 5.3 and 5.4. Consider first the explanatory game played from Galen to Harvey with respect to the functioning of the heart. Take first the different *rules of representation* used for the fourteen centuries until Harvey's own explanation and how they fared with respect to *accuracy*. Galen availed as a rule of representation the dissection of animals. It was only by Mondino in Bologna around 1315 that human dissection reappeared. It is clear that in a comparative evaluation of the two rules with respect to accuracy, the second rule is unambiguously better. The rules of representation had changed dramatically when Vesalius published his *Fabrica* in 1543. He collaborated with Jan Stefan van Calcar from the workshop of Titian to construct woodblock engravings of the drawings contained in the *Fabrica* – these anatomical drawings and illustrations reflected a naturalism in the depiction of human anatomy which was radically different from the conventional medieval drawings. Vesalius personally supervised the actual printing in Basel. With the use of printing technology, the drawings and diagrams could be copied and multiplied with accuracy and, thus, the representation of the explanandum phenomenon became more precise and more easily accessible. The direct representation by means of the senses in the process of dissection were unambiguously *less accurate* than the printed representation of the heart and the other organs by means of drawings and other illustrations.

Using these *rules of representation*, Harvey's main innovation was to introduce novel *rules of inference*: the analogy of the heart as a pump was, again, unambiguously better with respect to *consistency* in comparison with the analogy of the blood which ebbs and flows in the arteries employed by Galen (Acierno, 1994, p. 217). The introduction of a further novel rule of inference by Harvey was crucial: using simple mathematical calculations rather than ordinary language made the explanation of the *circulation of the blood* appear closer to the *truth* vis-à-vis other explanations offered up until then.

The rules of representation evolved further with the use of the microscope, first employed in the middle of the seventeenth century by Malpighi. These rules were unambiguously *more accurate* in depicting the anatomy of the heart. However, even if such a positive evaluation with respect to accuracy is reasonable, the institutional setting of the time permitted the following incident to take place: although Malpighi was a very modest and gentle man, the immense scope and impact of his microscopic studies provoked such envy that in 1648 his villa was burnt by adversaries; his papers, notes and manuscripts were destroyed; and his laboratory equipment was ruined. With respect to *freedom of expression* (which we will discuss in Chapter 6), the rules followed by his critics are to be evaluated as worse than the rules followed by the adversaries of Galen (though not of those followed by the adversaries of Miguel Servetus, who was sentenced to death by burning at stake in Geneva in 1553 with one of the apparently last copies of his book *Christianismi Restitutio*, which contained the accurate description of the small circulation of the blood chained to his leg!). In such cases, a multidimensional evaluation of different rules with respect to different properties becomes more complex, though by no means impossible: one is to proceed to a *more complex evaluative judgment* also involving the *weighing of the importance of the different properties*, both epistemic and non-epistemic.

5.4.2 Normative Appraisal of Interpretations

Consider now the interpretative problem regarding the Antikythera Mechanism. It consists, as we have seen, of (a) successfully reconstructing the representation-bearer; (b) successfully specifying the representational object, that is, what the device was supposed to represent at all; (c) inventing hypotheses on how exactly (a) and (b) are connected. The success in the reconstruction of the representation-bearer can be judged with respect to *accuracy*. For decades the identification of the nearly thirty gears of the Mechanism was made by using photographs and microscopes or basic X-rays – the high-resolution techniques Polynomial Texture Mapping and Microfocus X-ray Computed Tomography applied in 2005 offered an unambiguously *better reconstruction of the representation-bearer with respect to accuracy*.

What did the device represent at all, that is, what was the representational object? Consider the specification put forward by Price (1974) using basic X-rays that the representational object consisted of the planetary motions of the Sun, Moon and all five planets known in antiquity and compare it with the specification by Freeth et al. (2021, p. 12f.), which is *vastly more accurate*: "[The Antikythera Mechanism] calculated the ecliptic longitudes of the Moon, Sun and planets, the phase of the Moon; the Age of the Moon; the synodic phases of the planets; the excluded days of the Metonic Calendar; eclipses – possibilities, times, characteristics, years and seasons; the heliacal rising and settings of prominent stars and constellations; and the Olympiad cycle – an ancient Greek astronomical compendium of staggering ambition. It is the first known device that mechanized the predictions of scientific theories and it could have automated many of the calculations needed for its own design – the first steps to the mechanization of mathematics and science."

The third part of the interpretative problem concerns the ways that the representation-bearer is connected with the representational object. Consider the very first hypothesis put forward already in 1903 by the numismatologist Svoronos: the device was supposed to be an instrument of navigation, an astrolabe.[14] Consider next the hypothesis proposed by Kalligas (2016), which I have briefly presented in Section 5.3. Theon in his *On Mathematical Issues Useful for the Reading of Plato* mentions that he has himself constructed a mechanical device in the tradition of σφαιροποιία, adhering to Plato's requirement implicit in the phrase from Timaeus 40d1-3 according to which "to tell all this [sc. to describe the movements and the positions of the heavenly bodies] without using visible models would be a labor spent in vain."[15] Commenting directly on this passage, Pierre Duhem remarks (1908/2003, p. 16): "For Adrastus of Aphrodisias and for Theon of Smyrna, apparently also for Dercyllides, the mathematician should adopt an astronomical hypothesis that conforms to the nature of things. However, for these philosophers, this conformity is not to be evaluated by reference to the principles of physics adduced by Aristotle; it is ascertained by the possibility of constructing, by means of suitably encased solid spheres, a mechanism representing the celestial motions." According to Kalligas (2016, p. 187f.): "Theon's engagement with the making of a mechanical model of the cosmos certainly points back to a tradition of constructing similar instruments, some of which were said to have been the mark of that ingenious paragon of mechanical achievement, Archimedes (3rd cent. BC). The most striking report is the one by Cicero in his *De re publica* I 21–22 regarding two *sphaere* that the conqueror of Syracuse, Marcus C. Marcellus, had brought home from that city and were attributed to the art of Archimedes." Is such a hypothesis regarding the connection of representation-bearer and representational object true? Given the available evidence, the hypothesis that the Antikythera Mechanism (belonging to the species of instruments constructed since at least Theon of Smyrna) depicts the 'cosmos' is *closer to truth* than alternative hypotheses such as the original one that the device is an instrument of navigation, an astrolabe.

Closing, the case of text interpretation also discussed in Section 5.3 is not essentially different. Text interpretations, as reconstructions of nexuses of meaning, can be compared to each other and evaluated. Insofar an interpretation of a text can be closer to *truth* than an alternative one by virtue of offering more fitting reconstructions of the respective nexuses of meaning. However, *aesthetic values* might be deemed equally important for the evaluation of interpretations: some interpretations might be more beautiful than true. This is a common claim among literary theorists, a claim that is both plausible and correct: the same set of interpretations can be evaluated with respect to different values and found to be satisfying some of them while not satisfying others. As long as the critical discussion of the text interpreters appeals to empirical evidence and uses intelligible arguments in favour of some text interpretations vis-à-vis others with respect to truth, *hermeneutic objectivity*[16] can prevail and insofar text interpretation – contrary to the claim of deconstructivists and other postmodern authors – is to be viewed as a normal scientific activity.

In the case of the text interpretation of the passages on the invisible hand in the work of Adam Smith discussed in Section 5.3., the comparative evaluation of the three interpretations with respect to *truth* is possible. The standard interpretation is founded on solid empirical evidence from a variety of sources, since evidence is called upon not only from other similar passages of the same author but also from other authors and from the intellectual environment that Smith was part of. Given this evidence, the interpretation of MacFie which considers only three passages in which Smith mentions the invisible hand and not a broader array of empirical material makes it less acceptable than the standard interpretation. Let us finally consider Rothschild's interpretation of the invisible hand as an ironic device: (a) "One reason to suspect that Smith was not entirely enthusiastic about theories of the invisible hand is that these theories are condescending or contemptuous about the intentions of individual agents. Smith's three uses of the phrase have in common that the individuals concerned are quite undignified; they are silly polytheists, rapacious proprietors, disingenuous merchants" (1994, p. 320). Smith was known to be a defender of individual liberty and of the independence of individuals. For such an author, such an account of the invisible hand seems to be contradictory, if seriously meant. (b) Rothschild also stresses that Smith was in fact quite critical of established religion. The comments on religion in his opus, like those of Hume, are conscious of public opinion, but also ironical. Given that, the invisible hand should not be viewed as a way of expressing Smith's religious beliefs, that is, as the hand of the Christian deity (apparently consistent with deism). If one considers other passages in his work, one can find that they are also ironical rather than pious, as when he, for example, speaks of the "all-wise Being, who directs all the movements of nature; and who is determined, by his own unalterable perfections, to maintain in it, at all times, the greatest possible quantity of happiness" (Smith 1759/1976, p. 235). Rothschild's interpretation is, thus, supported by solid evidence based on the consideration of a large amount of relevant material. Insofar, as is common in science, the empirical evidence does not allow one in this case to make a decision and clearly favour one interpretative hypothesis. The stance to be adopted is to abstain from a judgment regarding which of the two hypotheses, the standard one or the ironic one, is closer to truth. The critical discussion as more evidence emerges might enable a more informed decision in the future.

5.5 THE METHODOLOGICAL UNITY OF SCIENCE AND HUMANITIES

Core epistemic activities include not only explanatory endeavours aiming at answering 'why?'-questions but also interpretative endeavours aiming at answering 'what is the case?'-questions. They are both species of epistemic problem-solving activities directed at different epistemological goals.[17] The outcomes of all such activities that are continuously produced during the ongoing interaction between epistemic agents are fallible and can be the object of a comparative evaluation with respect to different properties, epistemic and non-epistemic alike. The rationality of the process is manifested in the concrete informed decisions made by the epistemic agents during the process and by the appropriate institutions allowing for the possibility of criticism without negative effects to the critics.

Thus, both explanations and interpretations can be normatively appraised within the premises of a comparative approach using the same normative standards involving intersubjective intelligibility, testability with the use of evidence and rational argumentation. There are, of course, very different techniques that are employed in the diverse scientific and humanistic disciplines. But *there is not any in principle* difference between science and humanities constituting a dichotomy at the methodological level, if a comparative approach of normative appraisal is accepted and a view of procedural rationality is endorsed.

6 The Formal Institutions of Science

Science is indisputably one of the greatest and most astonishing cultural achievements of the human species. The successive endeavours of many generations of scientists have produced an increasingly accurate image of the natural and social world, an image constituted by a huge variety of empirically tested representations, in the form of theories, models and artefacts. Modern science is embedded in a social context without which it could not exist. Its characteristic form largely depends on a specific pattern of institutions that enable the activities of the working scientists and channel the scientific process. The scientific venture is, hence, a social process embedded in the institutional framework of the society consisting of informal institutions, such as *conventions*, *moral rules* and *social norms* and formal institutions, most importantly *legal rules*.

6.1 THE STATE AS THE ENFORCEMENT AGENCY OF FORMAL INSTITUTIONS

Scientific activities are conducted in territories that are controlled by states that have the power to enforce specific rules for all agents, individuals and organisations, using violence. It has taken many forms in the course of history depending on the kind and extent of resources that the rulers have had in their command and on a great array of historical contingencies that cannot be discussed here.[1] According to the classification of institutions that I have proposed in chapter 4, the state is the enforcement agency of the formal institutions of a society and what we will focus on here more specifically are the formal institutions that regulate science. In other words, what is of interest in our context is the sum of the legal rules intended to regulate the scientific process (Gascoigne, 2019).

There is a natural interest by all rulers to control the production of belief systems that are useful for attaining their primary aim that is to establish their rule in the territory that they control. It would be utterly naïve to suppose that there would be a natural tendency of state agents to allow epistemic problem-solving activities, if this would endanger their position of power. That violence has come to be exerted in a relatively orderly way allowing different kinds of freedom to be successively granted to ever more groups of subjects by state authorities has been a very slow historical process following different trajectories in the countries of the West where this process has been originally set in motion. As North, Wallis and Weingast rightly stress in their *Violence and Social Orders* (2009), it was first a *rule of law of elites* that kings and other rulers had to respect in return of support of their claim to rule, before the equal treatment under the law was extended to men with property, all men and finally women. There has been nothing deterministic in this process that could have unfolded entirely differently, if the historical trajectory had taken another path due to a variety of factors.

In the case of the period of the Scientific Revolution that I have briefly discussed in Chapter 4, some historians and social scientists stress the importance of the emergence of *competitive political structures* which led to the increase of individual freedom (Bernholz et al., 1998; Jones, 2003) and allowed the expression of critical ideas without pernicious consequences for the critic. Whatever the relevance of this factor has been historically, the most important point for our discussion is that the formal institutions of science emerge as an outcome of a process that scientists themselves either cannot influence at all (in the case of authoritarian political regimes) or only minimally (in the case of liberal political regimes). The arena of science is never entirely *auto*nomous in the sense that it can be regulated entirely by the rules that scientists themselves may impose upon their own epistemic problem-solving activities, but always *hetero*nomous, since those who control the exertion of force in a society have the final word about whether and to what extent such activities are permitted.

Classic examples of authoritarian political regimes are the cases of totalitarian Nazi Germany and Soviet Union. The formal rules that these regimes have imposed on their scientists have given rise to the well-studied cases of Nazi science and Soviet science. The state organs using violence or simply the threat of violence were able to influence the epistemic problem-solving activities in a dramatic way and so to determine their outcomes decisively. The so-called 'Lysenko affair' highlights the extent to which scientific autonomy can in fact be impaired by authoritarian regimes (Pollock, 2008, ch. 3). Trofim Denisovich Lysenko was an agronomist who rose to prominence in the former Soviet Union in the 1930s in a situation where the authorities wanted to increase the production of wheat after years of very poor harvests. Lysenko's experimental program promised to deliver a dramatic improvement of crop yields, if only the first generation of seeds were treated, since the effects of the treatment would be inherited. His proposal ignored the genetics that had been developed by Morgan in the wake of the rediscovery of Mendel – the concept of a gene being 'bourgeois idealism'. In other words, his proposal was at odds with classical genetics which consistently rejected the inheritance of acquired characteristics, that is, Lamarckism. Lysenko's ideas were endorsed by the communist party[2] and became official Soviet policy, and indeed official Soviet science whereas prominent scientists supporting orthodox genetics were denunciated, prosecuted and imprisoned. The meeting of the Lenin Academy of Agricultural Sciences in 1948 resulted in the adoption of Michurnism, an account of epigenetic inheritance commonly known as Lysenkoism, as the "only correct theory" to be taught in the USSR. Lysenko at the end of his report on problems with Mendelian genetics stated that "the Central Committee of the Party has examined my report and approved it" (Krementsov, 1997, p. 172).

In the case of liberal regimes, typically constitutional democracies of our time, freedoms to engage in scientific activities are secured one way or another to some degree. The government still takes an interest in influencing science in many ways, backed by more or less efficient administrations. The range of interventions can vary a lot between allowing only public organisations to conduct scientific activities, paying their members' salaries funded by taxing the constituency, to allowing also private organisations, often corporations, to conduct scientific activities funded by entirely private means. Modern states typically run ministries and specialised government agencies manned by public employees to support, influence or directly control scientific activities undertaken by thousands of specialists in their territory. Furthermore, international associations of states also issue regulations meant to influence scientific activities across borders. Those regulations differ fundamentally from national legislation because they lack a specialised enforcement agency, so that they often remain only somehow indicative, their ultimate enforcement being contingent upon the will of national governments. Governance of science has, thus, successively become very complex with a very specialised set of legislative rules aiming at regulating epistemic problem-solving activities.

In sum it should be clear that, *along with the informal institutions also formal institutions, i.e. legal rules, make up the institutional framework that regulates the arena of epistemic problem solving activities that we call science.* This institutional matrix defines the way that the scientific game is played. The complex patterns of scientific activities undertaken in everyday scientific life crucially depends on the prevailing institutional rules: other rules, other game. The relationship of the informal and formal institutions of science can be in principle of four kinds: they can either complement each other, they can substitute each other, they can be in conflict with each other or be entirely neutral to each other. This is due, to repeat, to the different kinds of processes that give rise to them in the first place. There is no reason to suppose that there must be a somehow beneficial institutional integration that will harmoniously regulate the unfolding of scientific activities. But once the institutions have taken some shape, they exert their effects while continuing to evolve.

6.2 THE GAME OF SCIENCE: COMPETITION AND COOPERATION

The scientific institutions prevailing at any moment of time structure the way that the game of science is played. The scientific actors, individuals and organisations, have different aims while pursuing their epistemic problem-solving activities, so that their social interaction is not always of a cooperative nature, but rather competitive interactions along some dimensions are bound to emerge. Scientific *competition* is an evolutionary process of trial and error among individual scientists and organisations pursuing many different aims varying from the search for truth to peer recognition and monetary rewards.[3] Scientific agents engage in problem-solving activities that include constant choices under conditions of scarcity. Competition for recognition among peers for scientific achievements like publications or experiments can only take place, if sufficient support for resources is provided for the problem-solving scientists. Insofar, the purposes of scientists are connected with the purposes of agents outside the arena of science, and this is the way by which science is connected with other social domains such as markets, politics, and religion. *Competition for peer recognition* is thus tied up with *competition for resources* (even more so, if one takes into consideration that scientific outcomes have to a great degree the quality of a public good in the economic sense, a complex problematic that cannot be tackled here).

The simultaneous effect of informal institutions, formal institutions and the competitive efforts of scientists shapes *an institutionally constrained scientific competition*. The arena that we have called science is both highly *competitive* (Albert, 2011; Albert, Schmidtchen, and Voigt, 2008) *and* highly *cooperative* (Wilholt, 2016; Heesen, 2017; Boyer-Kassem et al., 2018).[4] The specific mix of competition and cooperation depends on the concrete way that the institutional framework of science has been shaped. During their socialisation process, the individuals who later become scientists have learnt the scientific conventions, moral rules and scientific techniques of the scientific community they live in. When working in research centres or universities, the agents are already familiar with the legal rules, and they have learnt the degree to which the state protects or infringes their rights. Because they have gone through the same learning history, individuals and organisations largely share the same informal and formal institutions, that is, the rules of the game that make them the specific actors of the specific scientific game they are engaged in. They define the *cooperative element* of science.

The systematic integration of the institutional framework into the study of science (Jarvie, 2001) leads to a series of insights, the most important being that it offers the basis of determining the *content* of the scientific competitive process. Institutions determine what actions are permitted in the competition process, that is which parameters of scientific activity are allowed and which are not. Scientific agents are driven by the incentives rooted in the institutions to focus on those activities that are allowed. Hence, by allowing scientific agents only limited action parameters, the institutions channel their innovative potential to a specific direction. If the institutional structure allowed extensive genes manipulation of humans, for example, individual scientists and scientific organisations would invent respective sophisticated representations and suitable skills would be developed over time.

Beyond this, institutions determine the *speed* of the scientific competitive process. The dynamics of the competitive process depend on the payoffs, that is, the utility increase expected by the scientific agents engaging in specific scientific activities. The strength of these incentives, however, depends on the institutional framework, which can vary considerably. Patent laws, to take just one obvious example, regulate the rate of scientific inventions in any specific scientific domain, because they ensure that the patent owner will reap all possible monetary benefits from a scientific discovery.

The institutionally constrained competition determines, thus, *the blend of cooperation and competition among the participants in the game of science*. The metaphor of the invisible hand (Leonard, 2002, p. 143) is a way to describe the working of such an institutionally constrained competition. There is nothing that guarantees that such an invisible-hand process will give rise to emergent unintended outcomes which are in some sense "beneficial" (Hull, 1997). The opposite is very often the case – the emergent patterns might indeed be "pernicious."[5] The outcomes of the game of science are historically contingent (Kitcher, 1993, 2001). Their quality decisively depends on the intricate mix of informal and formal institutions and the way they structure the competitive efforts of the participating scientists and organisations. It is, thus, the specific *content* of the informal and formal institutions that structure the *competitive game* in an *appropriate* way that enable the *advancement* of science to take place.

6.3 THE INSTITUTIONALISATION OF CREATIVITY AND CRITICISM

As we have seen in Sections 6.1 and 6.2, the informal and formal institutions of modern science that have come to prevail in a long evolutionary process in the West are due to a historical contingency. On the one hand, specific informal institutions encapsulating *the scientific method* emerged during the Scientific Revolution. The emergence of *competitive political structures*, on the other hand, led to the increase of individual freedom and allowed the expression of critical ideas without pernicious consequences for the critic. An intricate mix of informal and formal institutions has come to prevail in most parts of the world in the modern era, which increased the *freedom of expression* which in turn gave rise to a *free competition of ideas*. The *organizational structures of modern universities* which have emerged in a gradual secularisation process have become decisive in pooling intellectual and material resources offering a secure platform for the generation and critical discussion of abstract theoretical and practical problems.

Science is, thus, embedded in these broader normative structures of the modern world. What distinguishes it from other social arenas, religion, for example, is the sophistication and systematicity by which empirical evidence is generated and assessed, something that is enabled by the social and cultural arrangements encapsulated in the institutional framework of science. It is the *possibility of criticism with the use of evidence provided by this framework*, which acts as a corrective to the error-prone problem-solving activities in which scientists, like ordinary people, are engaged; errors ranging from fallacious mental models that do not give an accurate representation of the environment to fallacious inferences (including confirmation biases, erroneous probabilistic calculations and much more). To the extent that this auto-correcting process of individual scientists and scientific organisations is enabled by the prevailing institutional framework, sustainable *scientific progress* takes place.

This claim does not hinge upon the concrete conceptualisation of scientific progress in terms of growth of knowledge, of truthlikeness or of problem-solving. I would like, thus, to *embrace ecumenism* with respect to the diverse notions of scientific progress. In other words, I would not like my argument to be held hostage to the details of some particularly nuanced version of scientific progress. The contemporary philosophical discussion on scientific progress revolves around four approaches to scientific progress, which deal primarily, if not exclusively, with the conceptual issue of *what is meant by scientific progress*, which is viewed as the most fundamental issue (Niiniluoto, 2023, p. 28): according to the *semantic approach*, scientific progress is conceptualised in terms of increasing truthlikeness (Niiniluoto, 2014, 2023; Oddie, 2014); according to the *epistemic approach*, scientific progress is conceptualised as the accumulation of knowledge (Bird, 2016, 2023); according to the *functional approach*, scientific progress is conceptualised as increasing functional effectiveness (Kuhn, 1962/1970, Laudan, 1977, 1981; Shan, 2019, 2023); finally, according to the *noetic approach*, scientific progress is conceptualised as increasing understanding (Dellsén, 2021, 2023).

All these definitions of scientific progress are important in themselves, and it is not my claim that they should all be rejected.

They offer alternative views on how scientific progress should be conceptualised adopting alternative epistemological approaches on how the structure of reality can be grasped. I hold the view that the *idea* of scientific progress is comprehensible without necessarily invoking a specific *criterion* of scientific progress or offering a formal *definition* of it. The focus of my claim is that the institutions of science permit scientists to circumvent their inherent cognitive limitations and to improve their activities by means of criticism. Scientific progress manifests itself in virtue of the creative invention and advancement of multiple ways to accurately represent reality and the critical appraisal and retainment of only those ways that have successfully passed empirical tests as we have seen in Chapters 4 and 5.

7 The Search for an Adequate Constitution

The social arena is replete with actors, individuals and organizations, which take influence on each other in myriad ways. Economic, religious, artistic, political, athletic and many other types of groups are in constant interaction in different social settings, and there are mutual influences in diverse ways. Important for the case of science has been the interaction between religious organizations and scientific actors, dramatically exemplified in the case of the trial and condemnation of Galileo Galilei by the Roman Inquisition in 1633, a tribunal of the Catholic Church, one of the most enduring organizations in human history. However, the state as "that human *Gemeinschaft*, who within a certain territory – this: the territory, belongs to the distinctive feature – successfully lays claim to the monopoly of legitimate physical force for itself," according to the definition of Max Weber (1922/1972, p. 29), is in a position to dramatically influence all domains of social life. This is due to the fundamental asymmetry in the possibility of using violence on the part of the state, a possibility that always becomes an actuality in one way or another. With respect to science that interests us here, the state is the organization that enforces the formal institutions of science, and insofar is in the position to influence decisively the scientific process.

However, a world state never existed in history, so that political orders have always been *local*. Legislation has been enforced by states that could control only a specific territory. The rules that comprise the scientific method are, in contrast, *universal*. Ever since the Scientific Revolution science has progressed despite the existence of very different political movements such as totalitarianism, fundamentalism, colonialism and political regimes such as kingdoms, empires, dictatorships and democracies. Why? Because of the prevalence of the informal institutions of science, the rules that scientists have been following in their daily work independently of the political regime that they happened to be born in and to live in. Since most of the states today have come to adopt a constitution as a general framework of government organization, it seems that an effective protection of science can take place at the constitutional level. The purpose of this chapter is to work out this view.

7.1 SELF-ENFORCING CONSTITUTIONS

Politeia was the term used in Ancient Greece to denote a constitution, a purely descriptive term referring to an established political order comprising "all the innumerable characteristics which determine that state's peculiar nature, and these include its whole economic and social texture as well as matters governmental in our narrower modern sense" (McIlwain, 1940, p. 21).[1] In his very influential account on constitutionalism, McIlwain (p. 13) claims that a constitution is fundamentally a "set of principles embodied in the institutions of a nation and neither external to these nor prior to them" rather than some formally adopted text. This reflects a long-standing view in political theory. A prominent example is Montesquieu (1748/1989) who thought that an equilibrium between order and liberty in a polity cannot be discovered once and for all and that each regime must determine its own form of constitutional government, the success of its constitution being dependent on the quality of its political culture, what he called 'the spirit of the laws'. The most important function of a constitution was to make sure that "power must check power by the arrangement of things" (book 1, ch. 4) – it was the English constitution which embodied this principle because it provided the distribution of power between the legislative, the executive, and the judicial.

The view that the constitution should be a *written text* fixing a permanent framework of government and providing the fundamental law of a regime was due to Madison and came into effect when the US Constitution was established. Political power is on this view subjected to the discipline of a text (Loughlin, 2022, p. 3). The American constitution came to rest its authority on two great pillars, representative government and checks and balances, that is, diverse institutional mechanisms constraining political power. Thomas Jefferson reminding Madison that "the earth belongs to the living and not to the dead" favoured a regular review of the Constitution and its renewal by every generation, so that citizens could reaffirm their consent (1789/1958, p. 395). Whatever the arguments presented in the context of the establishment of the US Constitution, it is important for our purposes that it has always been seen as a problem *how a written text can reflect constitutional reality*. Opting in the end for the institutional innovation of establishing a Constitutional Court was meant to allow for some kind of constitutional evolution, but only within the narrow premises of a judicial review.

This modern conception of the constitution as a *text* rather as an *evolving set of beliefs and practices* did not remain uncontested. The French counter-revolutionary thinker Joseph de Maistre (1794–1795/1965, p. 103) proclaimed that the belief that "a constitution can be made as a watchmaker makes a watch" ignores the limitations posed by the particular circumstances and that this was one of the greatest errors of Enlightenment, since "the constitution of a nation is never the product of deliberation" (p. 107).[2] Edmund Burke (1791/1962, p. 134) also stressed that the quality of a constitution depends on the customs and traditions of a country and using the same metaphor contended that "an ignorant man, who is not fool enough to meddle with his clock, is however sufficiently confident to think he can safely take to pieces, and put together at his pleasure, a moral machine of another guise, importance and complexity, composed of far other wheels, and springs, and balances, and counteracting and co-operating powers." As Loughlin (2022, p. 36) correctly stresses regarding these authors, "constitutions, they maintain, can be no more be made than language is made."

In contemporary political theory, Hayek (1960, 1973/1982, 1976/1982, 1979/1982) has argued along these lines stressing that the constitutional rules are the result of *a trial-and-error process* incorporating the wisdom of many generations, so that a written text is a mere reflection of a more complex structure of beliefs and practices evolving over time. Brennan and Buchanan (1985) have used the traditional *contractarian view* of constitution stressing the *Reason of Rules*[3] that any constitution reflects. A third view conceptualizes the constitutional contract as a *principal-agent relationship* between the members of society and the government and focuses on the design of a contract that makes sure that the agent behaves according to the wishes of the people; that this is extremely hard, has been stressed by Tullock (1987, p. 317f): "The view that the government can be bound by specific provisions is naïve. Something must enforce these provisions, and whatever enforces them is itself unbounded." The German jurist Ernst-Wolfgang Böckenförde (1967, p. 60) made essentially the same point in his influential claim that "[t]he liberal, secularized state draws its life from presuppositions that it cannot itself guarantee."

It seems, thus, that even if a constitution is a text, its normative authority cannot be understood and its effects cannot be researched, unless one pays the due attention to the underlying structure of the beliefs and social norms that effectuates its emergence, stability and change. A constitution conceptualized as a bundle of conventions rather than as a contract can better explain that the explicit consent of the contracting parties is not necessary and that a constitution can be stable as long as there is no serious opposition (Ordeshook, 1992). According to Hardin (1989, p. 119): "Establishing a constitution is a massive act of coordination that creates a convention that depends for its maintenance on its self-generating incentives and expectations."

In sum, a constitution is made up of formal rules constraining the agents representing the state in their actions as well as of the respective informal rules that citizens follow. The formal constitutional rules will regularly specify individual rights, define the organs that make up the state as well as the range of their authorities and outline procedures for making sub-constitutional rules.[4] These formal constitutional rules need to be enforced somehow, since there is no guarantee that the government will respect them. The most important enforcers remain the citizens[5] who are themselves carriers of a constitutional culture in the form of informal constitutional rules: conventions, moral rules and social norms which prescribe whether the behaviour of government is constitutional or not. Ultimately, in order to continue to exist, a constitution must be *self-enforcing*: the limits to government action laid down in the constitution require a sufficient number of citizens who are willing to support it. Since there is always a great range of opinions among citizens about the appropriate role of the state and what actions constitute a transgression of citizen rights, the *coordination* of diverse opinions and the construction of a consensus about a set of state actions that trigger citizens' reactions is essential. Those constitutions that are constructed according to the principle of the rule of law, for example, need to be supported by the appropriate civic culture, one that both opposes government transgressions and polices the state in a coordinated manner.[6]

7.2 THE FUNDAMENTAL *HETERO*NOMY OF SCIENCE

Although constitutions have existed for centuries, the speed at which new constitutions have been drafted in the last three decades, following the collapse of the Soviet Union in 1989 and the downfall of dictatorships in Latin America and to some degree in Asia and Africa has induced scholars such as Loughlin (2022, p. 16ff) to speak of the 'Age of Constitutionalism'. Constitutional discourse has been largely normative, but the Comparative Constitutions Project presents the most extensive endeavour to code *de jure* constitutional rules (Elkins et al., 2009).[7] This huge data set consists of a list of some 650 different variables according to which all national state constitutions since 1787 are coded on an annual basis. Theoretical accounts of constitutional change have also been offered, starting with the landmark work of Voigt (1999), highlighting how even formally unchanged sections of constitutions can be subject to different interpretations over time, so that *de facto* constitutional change takes place. Data trying to capture the *de facto* side of constitutions rather than the *de jure* side are provided by the Varieties of Democracy Project.[8] According to the latest Democracy Report 2023, the share of the world population living in autocracies has gone up from 46 per cent in 2012 to 72 per cent in 2022; according to the dataset by Geddes et al. (2014, p. 313) of all regimes shaped since the end of the Second World War, more than half of them were transitions from one kind of autocracy to another kind of autocracy. As far as I know, there is neither a theory nor empirical evidence on how constitutional rules affect the scientific process in different countries or types of regimes.

It should be obvious that the *auto*nomy of science is a mirage if it is supposed to state a requirement that the agents of science, individuals and organizations are to lay the constitutional rules in order to regulate their activities by themselves and for themselves. Science will remain ipso facto fundamentally *hetero*nomous, as long as it is conducted in a territory where the fundamental rules are laid by the agents of a state that controls violence – unless the agents and the scientists are identical, something that is largely only a theoretical possibility. The *scientific conventions*, the *moral rules* of science and the *scientific techniques*, can help maintain a domain where informal rules will provide decisive normative guidance to the participants to the game of science, but the *right* to pursue scientific activities can ultimately only be *granted* by the state. Science can never be *entirely* self-governed. The constitutional question is to determine the extent and specification of this right and the philosophical task consists in the provision of arguments to answer this question.

The issue is *whether* and to *what extent* the state should grant a specific protected domain to scientific agents – individuals and organizations – guaranteed by constitutional provisions and less general laws. In other words, the issue is about appropriately delineating a protected domain to science vis-á-vis other arenas of social activity like, for example, religion or alternative systems of public knowledge. This is a problem that *can be* and *is de facto* addressed in the form of a constitutional issue: what are the highest institutional principles in an organized polity that should regulate the functioning of science, so that all the values that we deem important are appropriately reflected and sufficiently traded-off? Suppose in a constitutional assembly that initiates a new constitution or a constitutional change there is a political will to protect science: how should one address this issue?[9]

7.3 DEFENDING THE COMPARATIVE APPROACH

Every discussion about the merits of an institutional change at any level has to take into account a very simple matter of fact: "we always start from here." Any change that is to be introduced has to take into consideration the existing scheme of things, the status quo. For whatever reasons, there is always a set of rules at work, the *institutional a priori*, regulating the behaviour of agents in the respective social domain. Moving from here to there involves showing the path from the given situation to the one deemed important and desirable. The same is the case with respect to the constitutional issue of how to protect science, if such a protection is deemed important.

One way to protect science is by working out an *ideal* and proposing ways to approximate this ideal. This is largely the case in the ideals of falsificationism (Popper),[10] of "a society of explorers" (Polanyi)[11] or of "a well-ordered science" (Kitcher).[12] However valuable such ideal approaches are as general guides for enabling good science, they are less operational than a *comparative approach* that I would like to present and defend here. Science takes place not only in free societies but also under authoritarian regimes. A comparative approach is applicable also in such regimes, even if the prevailing conditions there might be very far away from an ideal arrangement of the scientific enterprise. Moreover, the availability and acceptance of an ideal approach is neither sufficient nor necessary for making judgments about whether a constitutional rule x is better than a constitutional rule y. In other words, in order to compare the merits of a constitutional rule x vis-à-vis the merits of a constitutional rule y, one does not need to appeal to a different alternative *z* which is supposed to be the 'best' rule. To consider an analogy, the fact that a person regards *The School of Athens* as the best fresco in the world does not reveal how she would rank a Monet against a Rubens.[13]

The main claim of the comparative approach is that "good science" emerges when the appropriate formal and informal rules are in place. However, in an evolutionary world of permanent change, these rules are not set in stone but are rather bound to change. Different societies have evolved along different trajectories, and therefore working out an ideal for the appropriate place of science "in the society" is an impossible task to accomplish. Any kind of proposed arrangement must definitely undergo adaptation as societies continue to evolve and priorities change. That is the main reason why a comparative approach is much more operational than the provision of an ideal model, which supposedly applies to all times and under all conditions. The feasible task consists in the concrete specification of the constitutional rules that govern scientific activities, and this specification must proceed from an analysis of the prevailing situation without reference to an atemporal abstract ideal.

So, the more modest proposal that I would like to endorse is the critical analysis of the constitutional rules framing science with respect to different values. Such an analysis takes into consideration the fact that science is an evolutionary process unfolding within rules which are themselves changing. However, their rate of change is lower than the rate of change of the process which they are channelling. They are the 'relatively absolute absolutes' (Buchanan, 1989), which can be the target of evaluation. This *procedural* approach is the opposite of an *outcome* approach which attempts to evaluate directly concrete scientific activities either praising them or repudiating them. Instead of forming judgements about the goodness of specific scientific actions, it is possible, preferable and more operational to form judgements about the goodness of specific rules. Such an approach takes into account the social character of science unfolding within rules and at the same time honours its evolutionary, procedural character.

To provide an obvious example: throughout most of modern history, scientific activities have been organized either purely privately or as an ecclesiastical affair. Even in these early periods, the appropriate place of science in a polity was still a constitutional issue: even if there was no public spending on scientific activities nor publicly run organizations that hosted such activities, the need to take a stance towards these activities at the highest level remained intact. When the monopoly of organized violence by the state is used not only to tax citizens in order to finance scientific activities but also to run state-owned agencies where scientific activities are conducted, the need for constitutional arrangements takes a thoroughly different shape. Finally, when the commercialization of science re-enters a scene dominated for some time by public or semi-public organizational structures, constitutional arrangements reflecting more nuanced evaluations are necessary. Therefore, the search for an adequate constitution of science is a permanent task.

7.4 CONSTITUTIONAL DEMOCRACY: THE LIBERAL PARADOX

In the context of the contemporary democracies of the West, a major challenge that needs be faced has to do with successfully addressing what I would like to call the *paradox of liberal democracy*.[14] Modern representative democracies are typically *constitutional democracies*, in which only certain issues are subject to the operation of the majority principle. In other words, at any moment of time, people or their representatives can make collective decisions on a vast range of issues by following an appropriate voting procedure after a period of deliberation, but by no means on *all issues*. In fact, the issues that are regarded as most important are issues that are not subject to majority voting. Reintroducing slavery in the US, for example, is not something that can be voted for in a referendum or as part of the agenda of a political party in elections. A series of such issues are decided upon in critical junctures of the history of a state and the decisions make up part of a constitution.

Many constitutional provisions in the democracies of the West protect basic individual liberties, typically by securing the rights of the citizens from state invasion. The paradox of liberal democracy consists exactly in this tension. On the one hand, the majority principle seems to be desirable because it can sufficiently accommodate the preferences of the citizens and thus express public sovereignty. On the other hand, the principle of protecting the rights of individuals (and groups) restricts the sovereignty of the people: if the people and their representatives decide by majority voting to violate the rights, the constitution and the agents entrusted with its protection will prevent them. This inherent conflict in every constitutional democracy is arbitrated at the highest level by stipulating which issues are to be decided on majoritarian grounds by following specific procedures and which issues should not be subject to any kind of majority voting.

The paradox of liberal democracy takes a specific form with respect to science. When considering the appropriate governance of science, a major task beyond determining the *content* of the respective provisions using the comparative approach, is to determine which issues should *be anchored in the constitution* and which should be subject to *majority voting*. The freedom of expression and the freedom of scientific research are obvious cases of liberties to be anchored in the constitution for both epistemological and political reasons.[15] But other liberties which *prima facie* do not directly have to do with research are also relevant, for example, the right to private property. The US Constitution, for example, protects private property rights mainly through the Fifth Amendment's Takings or Just Compensation Clause: "nor shall private property be taken for public use without just compensation." Intellectual property rights and their protection are of obvious importance for scientific research, but other rights usually anchored in constitutions are also essential for science. Determining whether they should be at all anchored in a constitution is an important task. Naturally, ordinary legislation about science will remain an issue determined by majority voting, though even in constitutional democracies there is no guarantee that transient majorities will produce worthy outcomes.[16]

Besides, it is of eminent importance to stress that a constitutional democracy normally treats equally all kinds of belief systems and traditions. Cosmological, religious, astrological, magical and all possible belief systems that have been produced by informal institutions distinct from the ones that guide scientific research are equally protected. In a liberal constitutional democracy, all belief systems are treated equally either because everyone can exercise her right to accept, support and express whatever beliefs she wants in virtue of exercising her right of freedom of expression or in virtue of exercising other constitutional rights like the right to religious freedom. The ambiguous issue, indeed the issue that Feyerabend (1978) in his provocative *Science in a Free Society* has raised, is whether the state should especially *prime* science vis-à-vis other traditions, that is, whether science should not only be *protected*, but be *privileged*. Should tax payers finance scientific activities? If yes, why? Should teachers in schools teach *only* sciences such as physics chemistry etc.? "The power of the medical profession over every stage of our lives already exceeds the power once wielded by the Church. Almost all scientific subjects are compulsory subjects in our schools. While the parents of a six-year-old can decide to have him instructed in the rudiments of Protestantism, or in the rudiments of the Jewish faith, or to omit religious instruction altogether, they do not have similar freedom in the case of the sciences. Physics, astronomy, history *must* be learned; they cannot be replaced by magic, astrology, or by a study of legends" (p. 74).

Here the paradox of liberal democracy comes nicely to the fore. On what level should it be decided whether science should be *protected* along other belief systems or *privileged*? Should this be done at the constitutional level, that is, at a level very hard or impossible to change or should this be decided by the majority rule in a parliament or in a referendum? Feyerabend in his relativistic fervour to show that 'Reason', that is, scientific rationality, should not possess a monopoly in the public sphere as the Church used to hold in the past, but that all possible traditions should be treated equally, presupposes dogmatically that a liberal constitutional democracy is the best form of government. It is only because he uses the standards of a liberal democracy that a relativist like him can voice his criticism on scientific rationality and its products. But why should we assume that a liberal constitutional democracy is the last word in structuring government? The rule of counting votes rather than taking arms when solving political conflicts was *invented* in Ancient Greece and *reinvented* in its representative form in France and the US in modern times. This is also a *human* invention, extremely valuable to be sure, but we do not know whether human creativity will provide an even better mode of government in the future.

Other authors, without endorsing the relativism of Feyerabend, and indeed adopting the opposite epistemological stance, that is, that science is the only arena of human interaction that provides genuine, though imperfect knowledge of the natural and social world, stress the danger that scientists have more power and influence on democratic politics than they should. The most recent example of this view is Pamuk's (2021) *Politics and Expertise* who advocates democratic scrutiny that "should be directed toward exposing the values and assumptions of science to ensure that they do not restrict democratic decision-making to narrow, partial goals" (p. 55). The concrete institutional proposal is to enable 'science courts', beyond parliamentary scrutiny which is nowadays prevalent in liberal democracies: "I suggest an adversarial institution that can be initiated by citizens, and where competing experts make the case for different position on a policy question with a significant scientific component. A citizen jury [...] interrogates the experts, then deliberates and delivers a decision, evaluating the facts and their practical implications together. The adversary structure of my proposal is intended to expose the background assumptions, potential biases, and omissions in rival expert claims as well as clarify the levels of uncertainty. The separation of scientist advocates from citizen jurors avoids the difficulties of mutual deliberation under conditions of unequal authority, while placing citizens in the seat of judgment" (p. 100f.). This recent attempt of allegedly 'democratizing science' is defective for many reasons. First, what Pamuk (and others) have in mind, is scientific advice, which is not the same as *doing* science, but has to do with how to use scientific results for policy purposes. 'Science court' is certainly a misnomer for a way to structure scientific advice – this is no court which has any kind of authority. Second, to the extent that the existence of facts should be assessed, it is not only unclear but simply impossible that citizens without a year-long scientific training will be able to make the complex decisions that will allegedly find the truth that could not be found in the existing tight institutional context that characterizes scientific discourse. Finally, there remains the issue of motivating the *average* citizen to participate in these kinds of exercises, given the huge appeal that alternatives such as viewing athletic competitions, participating in religious activities or simply staying home seem to have – though this problem could be solved by making the participation obligatory by law.

One cannot stress enough how distinct the two arenas are, science and democratic politics. A scientific community is not a political community. The truth-finding enterprise has nothing to do with the majority rule: counting votes is not the method of approaching the truth. That along with many other things, *choices* are also prevalent in science should not be confused with the choices made in democratic politics. It is an entirely different institutional context within which choices are made in science and politics. As we have seen in detail in Chapters 4 and 5, the choices about the epistemic properties of representations made by scientists are *not arbitrary* and they are clearly directed at evaluations of epistemic products. They are value-laden since all choices are value-laden, but to call for 'democratic values' to be respected when such choices are being made is surely confusing. If by 'democratic values' the use of majority rule is meant, then it should be clear that the scientific method has nothing to do with majoritarianism.[17] If by 'democratic values' the values of the public or its representatives is meant (Schroeder, 2017, 2021), then it should be clear that it is impossible to pay any attention to them, because in a pluralist democracy, 'the public' has not only *diverse* but also *contradictory* values.[18]

This does not mean that the members of the scientific community enjoy any specific right to *influence* democratic politics disproportionally beyond the rights that emanate from their *citizenship* or are entitled to somehow *direct* democratic politics according to their own preferences. All scientists are citizens, and they have equal rights and obligations with other citizens engaging in other kinds of social activities. Insofar, the issues of how to structure *scientific policy advice*, how to structure *science communication* and how to structure *science education*, all issues on how the *idiosyncratic* domain of science should exert influence in a democratic polity are constitutional issues that should be debated and decided upon. As are the issues of whether and to which extent *government funding* of scientific activities should take place and according to which rules this should occur.

Turning the ideal of the 'democratization of science' into reality either leads to the corruption of the intricate informal institutions of science, if counting votes instead of applying the scientific method is the aim, or it is an ideal that is de facto impossible to approximate, since the values of the democratic public are partly contradictory. Rather than corrupting science with thoroughly inappropriate ideals from politics, the alternative is to guarantee unlimited, *open access to science*. In other words, an open science would secure access to every individual independently of gender, race, nationality, age or a specific background. The existence of such an open-access order would reflect *liberal constitutional legitimacy*. Limiting all possible barriers to entry to the domain of science, by introducing concrete mechanisms to accomplish this aim, would furthermore secure the *contestability* of all the epistemic products generated in this domain.[19] Contestability is generated and retained by keeping an open access to the arena we call science, so that everybody can in principle become a scientist and contest scientific results from within, as a member of a scientific community. This is the only feasible way to criticize scientific results not only effectively, but indeed at all. Of course, for purposes of *use of scientific knowledge*, contestability from outside science is also necessary and mandatory, but not about *what constitutes scientific knowledge*.

7.5 SCIENCE IN AUTHORITARIAN STATES

Turning very briefly to the case of authoritarian regimes, the governance of science in them will depend on the degree of control that the authoritarian ruler or group wishes to exert on scientific activities. This will in turn depend on the structure of the preferences of the ruler, the time horizon (which will be longer in case the ruling political party is powerful and, thus, able to constrain the ruler) and other factors.

However, authoritarian regimes have most often constitutions. They are not protecting individual rights like in the case of constitutional democracies or limiting the powers of government or immunizing a private sphere from public interference (Galston, 2011, p. 229). They do not rest on popular sovereignty as is clear in the case of the Iranian Constitution which in article 2 states: "The Islamic Republic is a system based on belief in: 1. the One God (as stated in the phrase 'There is no god except Allah'), His exclusive sovereignty and the right to legislate, and the necessity of submission to His commands; 2. Divine revelation and its fundamental role in setting forth the laws."[20] So, what is the role of constitutions in authoritarian regimes?

According to Voigt (2020, p. 28): "After all, a central function of constitutions is to put constraints on government, whereas autocrats seem to aim at being as unconstrained as possible. But precisely because autocrats are suspected of being unconstrained by any rules, they have a particularly hard time making their own premises credible. If this is a disadvantage – e.g., because people refrain from investing in that country – then a rational autocrat could wish to be constrained by constitutional rules that enhance his capacity to make credible commitments. In addition, no autocrat can truly rule as a single person but needs the support of a loyal group. [A] constitution can be thought of as a coordinating device for that group. [...] To the degree that the constitution collects and explicitly names a number of aspirations, it might serve to increase the autocrat's domestic legitimacy. Further, to improve his reputation internationally, an autocrat might have incentives to include a number of basic rights in the document. Formulated somewhat differently, constitutions in autocracies can have a commitment function, a coordination function, but can also serve as window dressing."

In any case, the existence of constitutional arrangements regulating science in an authoritarian regime will definitely be an improvement compared to the alternative situation of inexistence of such arrangements. The mere fact that rules are recorded in a written form in the most important legal document of a country offers more security to the scientists undertaking research. The content and enforcement mechanisms of constitutional legislation will be radically different than in the case of a democratic regime, but according to the comparative approach endorsed here, it will represent an improvement vis-à-vis the completely idiosyncratic wishes of an autocratic ruler or ruling party.

8 Five Principles for a Quasi-Autonomous Science

Since the agents of the state avail of the monopoly of violence in a territory, they are ipso facto in the position to lay down the formal constitutional rules. Scientific endeavours can take place only if the state grants the right to scientists and scientific organizations to pursue their activities – insofar there is a fundamental *heteronomy* of science to start with. But even if science can never be entirely self-governed the sum of what I have called informal institutions of science defines the range of its autonomy: these rules have largely emerged spontaneously in a long historical process, and they will continue to evolve over time during the interaction of the participating scientists. *If* the scientific representations of reality and the possibilities of technological interventions in the physical and social world based on such representations are valued positively, then a political community *may* grant the right to do science to its citizens. The concrete content of the constitutional rules enabling and protecting science will ultimately be decided in a constitutional dialogue depending on the constitutional culture and the political traditions of the respective community. The content and enforcement mechanisms of constitutional legislation will radically differ in democratic and authoritarian regimes, whereas the informal rules will be homogenous across regimes and states.

The philosophical task in which I would like to engage here is to work out some general principles that should be adopted, if science is valued positively and should be protected. These are principles for a quasi-autonomous science. Three of them are substantive and two procedural.

1. *Guaranteeing freedom of expression.* Making use of this freedom allows all kinds of opinions to be expressed including scientific opinions and so

enables science in the first place. It is the main prerequisite for publicly formulating and defending theoretical constructs and thus, for science to emerge. It has been formally anchored in the form of a constitutional right in diverse written documents, most prominently in the Declaration of the Rights of Man and the Citizen in the French Revolution of 1789 (Art. XI): "La libre communication des pensées et des opinions est un des droits les plus précieux de l'homme: tout citoyen peut donc parler, écrire, imprimer librement, sauf à répondre de l' abus de cette liberté dans les cas déterminés par la loi" and in the First Amendment to the United States Constitution: "Congress shall make no law respecting an establishment of religion, or prohibiting the free exercise thereof; or abridging the freedom of speech, or of the press; or the right of the people peaceably to assemble, and to petition the Government for a redress of grievances." The German Grundgesetz specifically guarantees the freedom of science (Art. 5.3.): "Kunst und Wissenschaft, Forschung und Lehre sind frei. Die Freiheit der Lehre entbindet nicht von der Treue zur Verfassung."

2. *Mutual rational control by critical discussion*. The principle of critical examination discussed in earlier chapters manifests itself in the case of science in the mutual rational control by critical discussion. The theoretical constructs created by scientists and scientific organizations are subject to an inter-subjective control and criticism by peers. The general practice of intersubjective criticism goes back to ancient Greek philosophy[1] and has been invented by the pre-Socratic thinkers; the rules of the scientific method emerged during the Scientific Revolution. The mutual rational control is enabled (a) by the creation and sharing of common means of representation, very often in the form of a common artificial language, so that *clarity in communication* is facilitated; (b) by the respect of some minimal *logical requirements of consistency*; and (c) by the recognition of *evidence* which is *publicly accessible* and repeatable like experiments, observations and measurements as the impartial arbiter of controversies (as opposed to private aesthetic or religious experience).[2]
3. *Appropriate steering of scientific competition*. The domain of science must be appropriately delineated vis-à-vis other domains such as religion, politics or markets. This can be accomplished by appropriately institutionalizing scientific competition. The blend of cooperation and competition among scientists and scientific organizations is brought about by the ways that the formal and informal rules structure

science vis-à-vis the domains of *religion*, the domain of *politics* or the domain of *markets*. There have always been cases of institutional conflict between science and these social domains. The domination of *religious* institutions has been prevalent throughout history in literally all cultures[3] conferring authority to interpreters of sacred signs or documents suppressing virtually every kind of competitive challenge by extra-religious agents – consider the authority of the Pope for centuries in the Western culture. The domination of *political* institutions can cause the expansion of political control in all social domains and, thus, also in science as in totalitarian regimes – consider the authority of the Führer and his Volk to be respected by Nazi science or the authority of the communist party in Soviet science. The domination of *economic* institutions can cause the expansion of the viewpoint of economic relevance on scientific endeavours[4] – consider the authority of businessmen in their capacity as members of steering committees of universities and research organizations.

Freedom of expression as the principle *allowing science to emerge*, mutual rational control by critical discussion as the main principle of *error elimination* and appropriate institutionalization of scientific competition as the main principle of *delineation of science vis-à-vis other social domains* are all substantive principles. They are to be complemented by two procedural principles that must be adopted in order to safeguard science.

4. *Open access to the scientific community*. The first procedural principle has to do with securing the *contestability* of all positions of power within the domain of science which emanate from the success in epistemic problem solving. Competition and criticism within the scientific process can be best secured, if an unlimited, open access to science is guaranteed. It is only when scientific agents have the feeling that all their epistemic actions are in principle contestable by potential competitors, that they will be less prone to abuse their epistemic position. Competition is a mechanism of disempowerment, and it can mainly be kept alive, if *nobody is excluded* from participating in the scientific game due to gender, race, nationality, age or a specific background. Contestability is generated and retained by keeping an open access to the arena we call science, so that everybody can in principle become a scientist and *contest scientific results from within*, as a member of a scientific community.

5. *Appropriately fitting formal and informal institutions*. The logically possible relationships between formal and informal rules are those of complementarity, substitution, conflict and indifference. We do not have an empirical theory of the relationship between formal and informal rules with respect to effectuating an advancement of science in the form of ever more accurate representations of aspects of reality. Such a fine-grained theory would provide a valuable basis of orientation, but it is certainly the case that any attempt to change the constitutional rules of science should aim at the appropriate fit of the rules. In some cases, it will be expedient to aim at complementarities between the constitutional text and the prevailing constitutional culture, in other cases at introducing novel formal rules in order to incrementally push out or weaken informal rules and in still others to aim at clearly separating domains of scientific activity to be exclusively regulated by formal rules from those to be exclusively regulated by informal rules. This procedural principle of fine-tuning is probably the hardest one to implement since it is about appropriately tying constitutional text with constitutional reality and thus turning a *de jure* constitution into a *de facto* constitution of science.[5] A key to successfully applying this principle is the knowledge of the working properties of the different kinds of rules and their effects on the performance of science. But we will still have to make fundamental choices. In making these choices reasonably, we must imaginatively create alternatives, deliberate on them and critically discuss them. In this process, we will make errors, but we can learn from our errors. This is all that we have. As in many other cases, this is our predicament.

Epilogue

The Constitution of Science Is Written in the Heart of the Scientists

The five principles proposed in Chapter 8 may act as a guide for securing a quasi-autonomous science. But it is important to stress that the range of its autonomy will ultimately depend on upholding the informal rules of science comprising its ethos and its methodology, and these rules must also be defended by the scientists themselves: an eternal vigilance on the part of the scientists is required,[1] a vigilance that can ultimately secure that attempts at domination from the executive arm of government, religious authorities or organized economic interests will remain contestable.

The quasi-autonomy of science is fostered further by the fact that science has become increasingly international, so that even if a state threatens or indeed succeeds in annihilating its autonomy, it will be unable to do so but in a limited territory. The de facto internationalization of science, itself a spontaneous development, is probably the most effective way to secure the quasi-autonomy of science, since it essentially primes the informal constitutional rules against the formal rules. Insofar one could agree with Pasteur: "Le savant a une patrie, la science n' en a pas." One may remain, thus, optimistic that even if cosmopolitanism might be an impossible ideal for a citizen and probably also for a scientist, science may remain cosmopolitan and to a large degree autonomous.

Protecting science means protecting the informal institutions of science, the tacit Constitution of Science written in the heart of the scientists.

Excursus

The Value-Free Ideal for Science

A "VALUE-FREE SCIENCE": THE GERMAN DEBATES *WERTURTEILSSTREIT* AND *POSITIVISMUSSTREIT*

Is there one value that is more important than the others? Truth perhaps? This was the suggestion of Max Weber in 1904, which provoked the famous debate on value judgments, the *Werturteilsstreit,*[1] in Germany. It is important to briefly reconstruct the context within which that debate took place. The *Kathedersozialisten*[2] were university professors who were very concerned with the miserable situation in which workers were living in the end of the nineteenth century and advocated that economics and the social sciences, *qua sciences,* should be normative. These university professors who were all public employees (since all German universities were public and exclusively funded by taxes) claimed that the social sciences should be normative and advocated appropriate normative solutions to the so-called social question (*soziale Frage*) when teaching "from the Chair." They were therefore characterized by their opponents as socialists ex cathedra, *Kathedersozialisten.* Max Weber defended the alternative position, that economists and other social scientists, *qua scientists,* should not use their university position to preach socialism, since in a democratic polity they could use other fora to propagate their political views.[3] Economics and the other social sciences – at the beginning of the twentieth century still at their infancy – should rather be pursued as empirical, not normative disciplines. The "value-free principle" (*Wertfreiheitsprinzip*)[4] is summarized by Max Weber as follows: "Empirical science can teach nobody, what he *ought,* but only what he *can* and – possibly – what he *wants.*" ("Eine empirische Wissenschaft vermag niemanden zu lehren, was er *soll,* sondern nur, was er *kann* und – unter Umständen – was er *will.*")[5]

As a minimal requirement, he expressed the view that professors in the university should communicate to their students when they were expressing *factual statements* and when they were expressing *value judgments* and they should try to separate them. Thus, Max Weber's most interesting philosophical claim was probably his defence of the *very possibility of linguistically separating* factual claims whose truth depend on states of affairs (*Tatsachen*) from value judgments which express a personal perspective on the world (*Weltanschauung*).[6] Such a separation is logically possible: if one *wants*, then one *can* follow it. Closely tied to this possibility claim is the most important claim for our purposes: *that the factual claims presuppose the acceptance of truth as a value, science should consist merely of factual claims and, thus, truth is the only acceptable value for science.*[7] Insofar, the notion of a "value-free" ideal is certainly misleading – science should not be "value-free." Indeed, it *cannot* be "value-free" since science is made by scientists who are human beings and thus act within an ineliminable normative dimension. It is rather the case that science should be guided only by one value, truth. So, the ideal propagated by Weber and by the many who followed him should be rather called the "truth-oriented" ideal of science.

This view was subsequently defended by Karl Popper and Hans Albert with new arguments during the legendary *Positivismusstreit*[8] in Germany, which was initiated at a conference in Tübingen in 1961. They were arguing against Theodor Adorno and Jürgen Habermas, respectively, who among many other claims regarding the autonomy of the social sciences and the possibility of somehow directly grasping the totality of society by the means of dialectic, forcefully questioned the separability of facts and values, both at the ontological and at the methodological levels. Habermas accused Popper of being a positivist on the grounds that he allegedly limited rationality to the domain of scientific research (including social scientific research), whereas the domain of praxis was left to a largely irrational decisionism.[9] The alternative that Habermas favoured was the collapse of the fact-values dichotomy arguing that standards and descriptions are both set dialectically and that the concept of truth as correspondence "is itself a standard that requires a critical justification."[10] Albert convincingly argued that the "dialectic cult of a comprehensive reason" does not offer solutions that are able to meet its own elusive standards, offering only adumbrations, hints and metaphors rather than arguments open to critical appraisal.[11]

Albert has shown how what he called a technological transformation of *factual statements* into *hypothetical imperatives* for practical purposes is possible – without the use of value judgments. Systems of scientific statements consisting merely of factual statements can be turned with the appropriate logical transformation to technological systems, which indicate the appropriate means of intervention in the natural and social world.[12] If a scientific finding establishes a relationship between two phenomena of the type "every time that A occurs, then B occurs," then, if for practical purposes we want B to emerge, we have only to create the conditions for the appearance of A. Take the case of the quantity theory of money, according to which the general price level of goods and services is proportional to the money supply in an economy – assuming the level of real output is constant and the velocity of money is constant. Suppose that we want to use the quantity theory of money for practical purposes. The statement "Every time the money supply increases in an economy, and as long as GDP and the velocity of money remain unchanged, inflation increases proportionally" can be transformed into the equivalent statement "If one wants to reduce inflation, then one has only to reduce money supply, as long as GDP and velocity of money remain constant."

The upshot of this discussion is that a science consisting merely of factual statements can, in the tradition of Max Weber, merely answer the question "What can we do?," not the question "What should we do?" A practical application of scientific results consisting merely of factual claims is possible, but it is not the task or obligation of scientists to proceed to this application, but rather the task of policymakers who are entitled to do so, at least in a democratic polity. Such an application certainly involves the employment of value judgments, but it is the task of citizens and policymakers to express these value judgments and defend them in the political arena – scientists can also participate in this political discourse, but *qua citizens, not scientists*. Adorno and Habermas, on the other hand, defended the opposite thesis: that it is the task of social scientists guided by an emancipatory epistemic interest (*emanzipatorisches Erkenntnisinteresse*[13]) to intervene in the social world and change it into a specific direction. There is, thus, according to the Frankfurt school of critical theory, a primacy of praxis and of the values guiding the practical life, that is, social, moral and political values; truth as correspondence is certainly *a value, definitely not the most important value for science* and should in any case be justified. Is there such a justification? This question is answered in Chapter 2.

B "VALUE-FREE SCIENCE": HENRI POINCARÉ'S AND PIERRE DUHEM'S ARGUMENTS IN FRANCE

The discussion on the role of values in science at the beginning of the twentieth century in France was less passionate than in Germany and unfolded in the different context of the natural sciences rather than that of the social sciences. Henri Poincaré has argued forcefully in favour of the separability of factual claims and value judgments in his characteristic sharp style: "Il ne peut pas y avoir de morale scientifique; mais il ne peut pas y avoir non plus de science immorale. Et la raison en est simple; c'est une raison, comment dirai-je? purement grammaticale. Si les prémisses d' un syllogisme sont toutes les deux à l' indicatif, la conclusion sera également à l'indicatif."[14] Thus, both Poincaré and Weber had similar views arguing within different contexts: the scientific statements themselves (even the ones that refer to the social world) *can* be formulated in such a way that they are not of a normative character.

Pierre Duhem was the first who introduced and illuminated the problem of "the undertermination of theory by evidence," or the "undertermination thesis," for short. According to this thesis, the evidence cannot prove the truth of the theory: there are alternative theories that are not simply equally well supported by any evidence we have but that would still be equally well supported *in principle* irrespective of the amount of evidence that we might be able to collect. Starting from an account of the experimental practice in physics, Duhem defended the view that it is simply impossible to collect "raw data" from experiments that could unambiguously lead to the acceptance or rejection of a theory by using solely the means of logic. In other words, the link between evidence and theory is not deductive. Experimental observation is "theory-laden,"[15] that is, there is always an interpretation of the data produced in an experiment for many reasons, the simplest being that the tools of measurement are constructed using a specific theoretical repertoire, so that all experimental data are "theory-impregnated."[16] Crucial experiments are impossible in physics (Duhem, 1906/1981, p. 285), and there is a de facto holism of theories and hypotheses at work when a test situation takes place.

Duhem's view is important for the contemporary discussion for three reasons. To start with, he was the first who consistently pointed at the lacuna between evidence and theories which cannot be filled by *algorithmic means*. In other words, there is nothing automatic, no formula, no calculus which can deliver the verdict about the truth and thus of the acceptability of a specific theory. Instead, a genuine "choix des hypotheses" is mandatory, a human choice that must be undertaken by scientific agents. This means that the history of the discipline with all the choices of the scientists of the past becomes important, since they offer the storage of solutions to the problem of choice, a treasure that has been accumulated by innumerable trial-and-error attempts of the past. Logic must be complemented by historical epistemology, which is the first important lesson of Duhem's work.[17]

The second major innovation has to do with invoking other epistemic considerations besides truth,[18] which can help making a reasonable choice among hypotheses: "Sans doute, entre des ceux theories logiquement équivalentes, le physicien choisira; mais les motifs qui dicteront son choix seront des considerations d'elégance, de simplicité, de commodité, des raisons de convenance essentiellement subjectives, contingents, variables avec les temps, les Écoles, les personnes" (Duhem, 1906/1981, p. 437).[19] *Theory choice is undertaken on the basis of a plurality of epistemic considerations*, because logic alone is a necessary, but not a sufficient, condition for such a choice.

Finally, the third major innovation concerns Duhem's proposal that theory choice ultimately depends on the availability of *bon sens* (1906/1981, p. 330): "[L]a pure logique n'est point la seule règle de nos jugements; certaines opinions, qui ne tombent point sous le coup du principe de contradiction, sont, toutefois, parfaitement déraisonnables; ces motifs qui ne découlent pas de la logique et qui, cependant, dirigent notre choix, ces 'raisons que la raison ne connaît pas', qui parlent à l'esprit de finesse et non à l'esprit géometrique, constituent ce qu'on appelle proprement le *bon sens*." He does not offer a full theory of *bon sens*, of course, but what we have here is an archetypal description of a specific problem-solving skill with respect to theory choice, a competence that seems to be closer to "knowledge-how" rather than "knowledge-what." By way of introducing this kind of competence, Duhem extends further the range of values that seem to be important in theory, to include non-epistemic values such as impartiality and loyalty.[20] The ethical dimension of *bon sens* seems to concern a specific stance of the scientist vis-à-vis the non-epistemic or moral values – the necessity that he remains neutral towards them when he works. This is certainly compatible with the defence of a "value-free ideal" of science: the role that the moral values should play is merely instrumental, aiming at securing that theory choice takes place only according to epistemic values.[21]

The discussion on the "value-free ideal" in France has shown the full complexity of the issue of the relationship between science and values. As long as one focuses on scientific theories as the main entities making up "science," the question of their truth is a perfectly legitimate question. Whether it should be the only value is also a reasonable question to ask. When, however, the scope of the entities making up "science" becomes wider to include experimental practices and data collection, then it becomes in principle possible to ask whether some *further epistemic properties*, such as simplicity, fruitfulness etc., should also be taken into consideration in order to evaluate what has now become "scientific activity" rather than "scientific outcomes." Once one is prepared to undertake the next step and further widen the scope of what constitutes science to include the web of relationships between scientists, then *non-epistemic values* such as integrity, impartiality and loyalty also become important. Acknowledging the existence and importance of epistemic properties other than truth and of non-epistemic values in scientific activity is not equivalent to abandoning the "value-free ideal," however – this is what the French discussion at the beginning of the twentieth century has shown. Supposing that truth is indeed a value, is truth the most important value? If yes, how is this justified? If no, how should the weighing between different values take place and who should undertake it? Let us now turn to a brief review of the Anglo-Saxon discussion.

C "VALUE-FREE SCIENCE": THE ANGLO-AMERICAN DISCUSSION

Quine has famously updated Duhem's position in his *Two Dogmas of Empiricism* (1951/1980, p. 42): "The totality of our so-called knowledge or beliefs, from the most casual matters of geography and history to the profoundest laws of atomic physics or even of pure mathematics and logic, is a man-made fabric that impinges on experience only along the edges. Or, to change the figure, total science is like a field of force whose boundary conditions are experience. A conflict with experience at the periphery occasions readjustments in the interior of the field. [...] But the total field is so underdetermined by its boundary conditions, experience, that there is much latitude of choice as to what statements to reevaluate in the light of any single contrary experience. No particular experiences are linked with any particular statements in the interior of the field, except indirectly through considerations of equilibrium affecting the field as a whole."

Although the "Duhem-Quine thesis" clearly left much "latitude of choice" and, thus, opened the door of a series of values to enter the field of theory assessment, logical positivism was for a great part of the twentieth century the philosophical orthodoxy in the Anglo-Saxon world. Hans Reichenbach argued that knowledge and ethics are fundamentally distinct enterprises: knowledge is purely descriptive, ethics is purely normative. Therefore "[t]he modern analysis of knowledge makes a cognitive ethics impossible" (1951, p. 277).

Although predominant, the logical positivistic claims did not remain unchallenged. Rudner (1953, p. 2) stated for example: "In stating a hypothesis the scientist must make the decision that the evidence is sufficiently strong or that the probability is sufficiently high to warrant the acceptance of the hypothesis. Obviously our decision regarding the evidence and respecting how strong is 'strong enough,' is going to be a function of the *importance*, in the typically ethical sense, of making a mistake in accepting or rejecting the hypothesis [...] *How sure we need to be before we accept a hypothesis will depend on how serious a mistake would be.*"

Isaac Levi has broadened the discussion by suggesting that canons of inference limit the values scientists can and should use in their evaluation of scientific hypotheses. He has defended the view that, even if truth is not the sole operative value of science, *epistemic values* still make for sufficient criteria with respect to scientific thinking.[22]

In the sixties, the debate on inductive risk has also focussed merely on the *epistemic considerations* that are to be applied in the acceptance and rejection of hypotheses.[23] Carl Hempel (1965, p. 92) succinctly stated the problem of inductive risk: "Such acceptance (of a scientific law) carries with it the 'inductive risk' that the presumptive law may not hold in full generality and that future evidence may lead scientists to modify or abandon it." He stressed, however, that there was a specific goal of science which mandates the exclusive reliance on epistemic values.[24]

The discussion on the "value-free ideal" in the Anglo-Saxon world (Scriven, 1972) was largely centred around the view of science consisting of "the statements representing scientific knowledge"

(Hempel, 1965, p. 91), but there were also prominent voices who stressed the embeddedness of science in society, thus viewing science as a social enterprise rather than exclusively as a set of statements. Robert Merton (1942/1973) has elaborated on the four components of the "ethos of science" in his essay on *The Normative Structure of Science*: universalism, organized scepticism, communalism, and disinterestedness. Popper (1944/1957, p. 154) was probably the first that viewed science as a social enterprise but also stressed that "[s]cientific method itself has social aspects. Science, and more specifically scientific progress, are the results not of isolated efforts but of the free competition of thought. For science needs ever more competition between hypotheses and ever more rigorous tests. And the competing hypotheses need personal representation, as it were: they need advocates, they need a jury, and even a public. This personal representation must be institutionally organized if we wish to ensure that it works."[25]

Nagel in the *Structure of Science* (1961) defended the Weberian thesis of the possibility of distinguishing between factual and value judgments, notably in the social sciences,[26] and stressed "the self-correcting mechanisms of science as a social enterprise" that may decisively limit, though not completely eradicate unconscious bias and tacit value orientations of individual scientists.[27]

Kuhn's work provided an immense challenge of a series of logical positivistic doctrines, but I will focus on two main claims that are directly relevant to the issues that concern us here. First, the claim "of the unparalleled insulation of mature scientific communities from the demands of the laity and of everyday life. That insulation has never been complete – we are now discussing matters of degree. Nevertheless, there are no other professional communities in which individual creative work is so exclusively addressed to and evaluated by other members of the profession" (Kuhn, 1962/1970, p. 164). Second, the claim that "[t]here is no neutral algorithm for theory choice"[28] and that there are five standard criteria, that is, accuracy, consistency, broad scope, simplicity and fruitfulness for evaluating the adequacy of a theory.[29]

Lakatos by productively absorbing some of Kuhn's ideas reconstructed history of science as a history of competing research programmes providing a vigorous restatement and extension of Popper's critical rationalism introducing two central elements important for our discussion: *theoretical pluralism* and *competition*. At every moment of time, there is a *plurality* of research programmes consisting of a "hard core" at which a negative heuristic forbids us to direct the modus tollens at "this hard core" and a protective belt of auxiliary hypotheses to which we must redirect the modus tollens (Lakatos, 1970, p. 133). And there is a *competition* between the research programmes. A choice between research programmes becomes mandatory, and there are objective reasons to reject a programme (p. 155): "*Can there be any objective* (as opposed to socio-psychological) *reason to reject a programme, that is, to eliminate its hard core and its programme for constructing protective belts?* Our answer, in outline, is that such an objective reason is provided by a rival research programme which explains the previous success of its rival and supersedes it by a further display of *heuristic power*."[30]

Feyerabend's anarchic philosophy of science constituted a direct attack on the objectivity of such reasons, and most importantly for our purposes, questioned apart from the rationality of the scientific enterprise, the legitimacy of the claim to monopoly that science seems to raise in a democratic society.[31] Paul Feyerabend (1978, p. 96) was among the first to reflect and provide a concrete proposal on the role of science in a democratic society that favours the control of the judgments of scientists by elected committees of laymen: "(I)t would not only be foolish *but downright irresponsible* to accept the judgment of scientists and physicians without further examination. If the matter is important, either to a small group or to society as a whole, *then this judgment must be subjected to the most painstaking scrutiny*. Duly elected committees of laymen must examine whether the theory of evolution is really as well established as biologists want us to believe, whether being established in their sense settles the matter, and whether it should replace other views in schools."[32]

Following the lead of Bloor (1976/1991), Latour and Woolgar (1979/1986) or Rorty (1979/2018) relativistic tendencies became prevalent, so that the exact opposite of logical positivism with respect to the primacy of truth as value guiding the scientific enterprise has been seriously contended and discussed.[33] A controversy following up what seemed to become an increasingly serious attack on the authority of science by "social constructivists" (who suggested that scientific knowledge was socially constructed and thus should be treated on a par with other knowledge claims such as mythology, folklore, etc.) was of such an intensity that has come to be referred to as the "Science Wars."[34] What is important for our purposes is that during this debate, the epistemic authority of science has been seriously questioned but also that the normative standards for judging the quality of work in the humanities has been found deficient in comparison to the epistemic values that prevail in the natural sciences.

Arguing against prevalent relativistic tendencies, Larry Laudan (1984) developed an account that accommodated the presence of both agreement and disagreement in science. Scientific debates can be brought to a cognitive closure and a rational assessment of theories with respect to cognitive aims is possible.[35] He defended what he called the *reticulated model of scientific rationality* according to which (p. 62) "our factual beliefs drastically shape our views about which sorts of methods are viable, and about which sorts of methods do in fact promote which sorts of aims. [...] [T]here is a complex process of mutual adjustment and mutual justification going on among all three levels of scientific commitment."[36]

Helen Longino's attempt to present *Science as Social Knowledge* (1990) generalized the feminist critique of mainstream philosophy of science to include a view of an appropriately structured social process that leads to scientific knowledge – she calls this view of scientific knowledge *contextual empiricism* (p. 219). Longino denies that there is a genuine dilemma between science delivering truth and objectivity on the one side or merely a set of beliefs on a par with mythology, folklore, etc. Objectivity comes instead by degrees to the extent that hypotheses run through transformative criticism in a scientific community satisfying four criteria: (a) recognized avenues of criticism, (b) shared standards, (c) community response and (d) equality of intellectual authority (pp. 76–78). Truth cannot be the outcome of such a process of transformative criticism, but only consensus. "To say that a theory or hypothesis was accepted on the basis of objective methods does not entitle us to say it is true but rather that it reflects the critically achieved consensus of the community. In the absence of some form of privileged access to transempirical (unobservable phenomena) it's not clear we should hope for anything better" (p. 79).[37] Longino (2002, p. 128ff) elaborates further these criteria that she thinks are necessary to assume the effectiveness of discursive interactions. But what she calls "critical contextual criticism" is untenable, because it refers to an idealized epistemic community, a position identical with the one proposed by Habermas whose serious weaknesses are convincingly shown by Herbert Keuth already in 1993, but not taken into account in the Anglo-Saxon discussion.[38] Her main claim is that "the social position or economic power of an individual or group in a community ought not determine who or what perspectives are taken seriously in that community. Where consensus exists, it must be the result not just of the exercise of political or economic power, or of the exclusion of dissecting perspectives, but a result of critical dialogue in which all relevant perspectives are represented" (2002, p. 131). This requirement is of an utopian character, not applicable in any world inhabited by real human beings. As I am trying to show later in the book, *critical dialogue* is indeed a necessary condition for scientific knowledge to emerge – this is the old and important Popperian idea – but the task is to *foster existing institutions or to invent new ones* which will permit its effects to unfold in the world inhabited by real human beings.

Philip Kitcher (1993, 2001, 2011) honouring the social dimensions of science, but at the same time defending a more robust account of scientific knowledge enabling *significant* truths uses the map analogy to show "how maps designed for different purposes pick out different entities within a region or depict those entities rather differently" (2001, p. 56). Although the construction of an ideal atlas is impossible, "[t]he maps actually produced in human history are a selection of sheets from this atlas" (2001, p. 60). Kitcher acknowledges the diversity of human interests and illuminates how the sciences are directed at finding significant truths. He also shows how this is supposed to take place in a democratic society by developing the ideal of a *well-ordered science*: "science is well-ordered when its specification of the problems to be pursued would be endorsed by an ideal conversation, embodying all human points of view, under conditions of mutual engagement" Kitcher (2011, p. 106).

The point of departure is that modern science was not planned (Kitcher, 2011, p. 98) but rather a historical contingency grown out of the initial impulse of the members of the Royal Society, who were essentially members of a club, that is "gentlemen, free and unconfin'd" as they described themselves. The task is to find the appropriate locus of science as a prominent part of the system of public knowledge with respect to the ways it contributes to and is constrained by the goals of democracy (Kitcher, 2011, p. 86). The general issue is, thus, how to embed a spontaneously emerged system of public knowledge within the fundamental rules of a polity organized as a democratic order. There are two ways to address this issue, which is fundamentally about the governance of science. The first is by means of constructing an ideal and of proposing appropriate ways to realize the ideal. This is the way Kitcher favours: well-ordered science is an ideal, something at which our practices should aim and the key is to identify procedures for attaining or approximating the ideal. Kitcher proposes a series of alternatives that could enable citizens to engage in discussions with scientists in order to jointly determine what would be significant projects to pursue and the resources that should be devoted to them. A real-world deliberation is a way to approximate the ideal deliberation required by well-ordered science. He contends (2011, p. 125) that "without some understanding of where you want to go, efforts to improve on the status quo will be leaps in the dark." Kitcher is in the process of revising his views in his *Moral Progress* (2021), and his new ideas about progress in science have now taken shape in Kitcher (2023).

The ongoing debate about the "value-free ideal" partly elaborates on older arguments showing their interconnections with other discussions in philosophy of science and partly resolves around new issues. Putnam's (2002) forceful argument for *The Collapse of the Fact/Value Dichotomy* does not deny the *distinction* between facts and values, but only that nothing metaphysical follows from the existence of a fact/value distinction in a modest sense (p. 19). He stresses that "theory selection always presupposes values"[39] (p. 31) and that the role of epistemic values consists in telling us that we have arrived at truth (p. 32): "The claim that on the whole we come closer to truth about the world by choosing theories that exhibit simplicity, coherence, past predictive success, and so on, and even the claim that we have made more successful predictions than we would have been able to obtain by relying on Jerry Falwell, or on imams, or on ultraorthodox rabbis, or simply relying on the authority of tradition, or the authority of some Marxist-Leninist Party, are themselves complex empirical hypotheses that we choose (or which those of us who do choose them choose) because we have been guided by the very values in question in our reflections upon records and testimonies concerning past inquiries – not, of course, all the stories and myths that there are in the world about the past, but the records and testimonies that we have good reason to trust *by these very criteria of 'good reason'*."

One specific avenue that Putnam explored in this context is the problematic of thick concepts[40] in which a purely descriptive component and a purely evaluative component co-exist and are not 'factorable', that is, cannot be neatly distinguished (2002, p. 36). This 'entanglement' of the factual and the evaluative at the very level of concepts is somehow dramatized by Dupré (2007) who claims that this makes a value-free social science impossible. This criticism has been previewed by Max Weber (1917/1985) and has been convincingly refuted by Keuth (1989) whose work, however, is in German and has therefore not influenced the English-speaking discussion.[41]

Accepting the dramatization of the problematic of thick concepts must lead to the conclusion that there is no essential difference between political philosophy and social science in the end: all or nearly all concepts used in the social sciences are thick concepts. However, *scientific concepts* are *deliberately* created or modified in order to be used in the construction of a social scientific theory aiming at explaining aspects of the social world. One *can*, thus, deliberately choose to use concepts that are not thick for such theory construction, so that this overdramatized problem can be solved (an issue standardly treated in most textbooks on methodology of the social sciences). That these choices also involve values is evident, of course, but the truth content of the statements need not itself be normative. To present the problematic with the help of an example: is there really no difference between the claim "The inflation rate in Turkey in 2023 is 50%" and the claim "The central bank of Turkey must lower the inflation rate below 80%," because the term "inflation" is a thick concept?

Finally, the old underdetermination thesis is still being discussed, but the debate is rather on its extent, that is, whether it is transient, permanent or global (Biddle, 2013, p. 125; Brown, 2013, p. 831). Norton (2008, p. 18f.) claims that "the underdetermination thesis is little more than speculation" and that the question of the empirical underdetermination of scientific hypotheses should be decided on a case-by-case basis and Henschen (2021) is in agreement. A lot of recent work is on the further elaboration of the argument from inductive risk already mentioned above which was reintroduced in the discussion by Heather Douglas (2000) and in her *Science, Policy, and the Value-Free Ideal* (2009).[42] As is clear from the arguments discussed in the main text of the book, the debate about the Value-Free Ideal of Science is still as vibrant as ever.[43]

Notes

2 SCIENCE AND VALUES

1. For a review of the discussions on the Value-Free Ideal for Science in Germany, France and the Anglo-Saxon world, the reader should consult the Excursus to this chapter. Here I include only an extremely condensed summary of what is contained in the Excursus.
2. Steel (2010, p. 15) makes a further distinction between what he calls "intrinsic and extrinsic epistemic values. An epistemic value is intrinsic if manifesting that value constitutes an attainment of or is necessary for truth. For example, empirical adequacy is an intrinsic epistemic value because an empirically accurate theory is a theory whose consequences for observable phenomena are mostly true. In contrast, testability is not an intrinsic epistemic value, but a preference for testable hypotheses arguably promotes the attainment of truth indirectly by enhancing the efficiency of scientific inquiry (Popper, 1963). Thus, testability is an extrinsic epistemic value." Some authors adopt this distinction as, for example, Rolin (2015, p. 159) who goes so far to claim that moral and social values are extrinsic epistemic values: "For a value to promote the attainment of truth may mean that it leads scientists to support social arrangements that are instrumental in the epistemic success of science. For example, diversity is an extrinsic epistemic value insofar as it leads scientists to cultivate a diversity of perspectives, and this in turn facilitates transformative criticism in scientific communities (Longino, 2002, p. 131). While moral and social values are not epistemic values intrinsically, they can be argued to be extrinsic epistemic values on the grounds that they lead scientists to act in ways that are conducive to truth."
3. Kitcher (2001, p. 31) makes a distinction between transient and permanent underdetermination: "The underdetermination thesis obtains its bite when permanent underdetermination is taken to be rampant. [...] For in the mundane cases of transient underdetermination

scientists do resolve the dispute between alternative hypotheses, opting for one and rejecting its rival(s). According to the global underdetermination thesis, there is a way of developing the rejected rival(s) to obtain a theory (theories) that would be just as well supported by the new evidence – the evidence that allegedly puts an end to debate – as the doctrine that is actually accepted. Scientists thus make choices when there is no evidential basis for doing so. How do they break the ties? At this point, critics of the ideal of objectivity often insert their own psychosocial explanations. Since there are no objective standards for judging the victorious hypothesis to be superior, the decision in its favor must be based on values: scientists (tacitly or explicitly) arrive at their verdict by considering what fits best with their view of the good or the beautiful or what will bring them happiness." See also the useful discussion by Biddle (2013) who speaks of the "ideal of epistemic purity" in this context: "[T]he prescription to follow the evidence as far as it leads is shared by almost all critics of the ideal of epistemic purity. While there are few who appeal to arguments from underdetermination to support a radical relativism, according to which nature provides virtually no constraints on scientific reasoning, most philosophically-oriented critics do not. They argue that, although logic and evidence do not determine theory choice uniquely, they constrain it significantly; the influence of contextual factors should be minimized, so as to allow evidential considerations as decisive a role as possible."

4. Brown (2013, p. 834) calls this "the lexical priority of evidence over values. The premise of lexical priority guarantees that even in value-laden science, values do not compete with evidence when the two conflict. This is often defended as an important guarantor of the objectivity of the science in question." He follows, however, Anderson (2004) by claiming that the real problem is not the insertion of values, but dogmatism about values, and thus rejects such a lexical priority. He argues in favour of an account "in which values and evidence are treated as mutually necessary, functionally differentiated, and rationally revisable components of certification" (p. 838). Douglas (2009, p. 96) emphasises that values should never play the direct role of accepting or rejecting a hypothesis but may legitimately play an indirect role when "the values do not compete with or supplant evidence, but rather determine the importance of the inductive gaps left by evidence." According to Wilholt (2009, p. 96): "the protracted debate on science

and values has shown that it is deeply problematic to try to separate epistemic from non-epistemic, or cognitive from non-cognitive values" and contends that the problem of inductive risk is pervasive in science. Steele (2012, p. 899) analysing the role of scientists qua policy advisors claims that they "must often, for pragmatic reasons, deliver advice to policy makers in terms of a standardized evidence/plausibility scale that is cruder than the scale appropriate for representing their beliefs. In other words, the structure of the beliefs of scientists is often more complex than the predefined evidence scales, and there is no canonical projection. In these situations, scientists cannot avoid making value judgments, at least implicitly, when deciding how to match their beliefs to the required scale." Hicks (2014, p. 3290) claims that the "lexical priority of truth" should be followed by the qualification "from the scientific perspective." His "criticisms of this priority can be understood as recognizing that other groups of people, such as the pharmaceutical industry, have their own constitutive values, and from their perspectives the various values of scientific research are contextual. Specifically, truth is of paramount importance from the scientific perspective, yet it is much less important than the constitutive value of profit from the pharmaceutical industry perspective." De Melo-Martín and Intemann (2016) challenge the view that the argument from inductive risk actually undermines the value-free ideal, since it presupposes an assumption, which is, according to their view, problematic with such an ideal: that contextual values cannot legitimately play evidentiary roles. Hudson (2016) also challenges what he calls 'the uncertainty argument' which is essentially the argument from inductive risk (along with what he calls 'the moral argument') and defends the value-free ideal of science. Parker and Winsberg (2018) focus on modelling practices, stressing that purposes and priorities in modelling reflect interests, such interests, in turn, reflect values which are not just epistemic but also social (p. 128): "Just to be clear: our claim is not that modeling studies do not have epistemic goals; clearly they do. Our point is merely that the epistemic goals of modeling studies often stem in part from non-epistemic interests and values." Contessa (2021) discusses two broad approaches to the mitigation of inductive risk, which she calls respectively the individualistic and the socialised approach. Henschen (2021) asks "how strong is the argument from inductive risk?" and argues that it is weaker than commonly considered: Jeffrey's (1956) objection that the genuine task of the scientist is to assign probabilities and not to accept hypotheses and Levi's (1962) point that scientists need only to decide what to believe and not how to act have not been refuted by any arguments in the literature (at least not yet). Dressel (2022) also doubts whether the argument from inductive risk really refutes value freedom. Hoyningen-Huene (2023, p. 17), on the contrary, stresses that non-epistemic values enter also the decision about the performance of experiments in pure science and "the fact that scientific hypotheses are not only 'applied' outside of science, where 'the policy makers' have their say, but also inside science, and this on an absolutely regular basis." Magnus (2022) thinks that inductive risk is present every time someone forms a belief.

5. This is the classic enumeration of epistemic values by Kuhn in his "Objectivity, Value Judgment and Theory Choice" printed in his *The Essential Tension* (1977, p. 321ff.).
6. See Popper (1972), Hull (1988) and Longino (1990, 2002).
7. See Resnik and Elliott (2019).
8. As Sextus Empiricus in Book 1 of the *Outlines of Pyrrhonism* observes (164–169): "Οἱ δὲ νεώτεροι σκεπτικοὶ παραδιδόασι τρόπους τῆς ἐποχῆς πέντε τούσδε, πρῶτον τὸν ἀπὸ τῆς διαφωνίας, δεύτερον τὸν εἰς ἄπειρον ἐκβάλλοντα, τρίτον τὸν ἀπὸ τοῦ πρός τι, τέταρτον τὸν ὑποθετικόν, πέμπτον τὸν διάλληλον." ("The later Sceptics hand down Five Modes leading to suspension, namely these: the first based on discrepancy, the second on regress ad infinitum, the third on relativity, the fourth on hypothesis, the fifth of circular reasoning.")
9. This discussion mainly took place in Germany with respect to the possibility of "Letztbegründung" of ethics. See Apel (1973, 1976), Albert (1975, 1982), Kuhlmann (1985, 2010), and Keuth (1989, 1993). See also Grundmann (2017, p. 317ff.). There has also been some use of Agrippa's modes in Anglo-Saxon philosophy of science, see, e.g., Chakravartty (2017, p. 237ff.).
10. The Baron of Münchhausen was a German nobleman who became famous in the eighteenth century as a storyteller: in one of his stories, he managed to get himself and the horse on which he was sitting out of a swamp by pulling his own hair. As far as I know Gottlob Frege was the first to refer to Münchhausen in this context. See the preface of his *Grundgesetze der Arithmetik* (1893, p. XIX): "Wo ist den hier

der eigentliche Urboden, auf dem Alles ruht? Oder ist es wie bei Münchhausen, der sich am eigenen Schopfe aus dem Sumpfe zog?" Karl Popper in the *Logik der Forschung* (1934/2003) offered a similar discussion with reference to Jacob Fries's *Neue oder anthropologische Kritik der Vernunft* (1828–31), which he baptised as "Fries's Trilemma." In the English translation, that is, the *Logic of Scientific Discovery* he writes (§25, p. 75): "The problem of the basis of experience has troubled few thinkers so deeply as Fries. He taught that, if the statements of science are not to be accepted dogmatically, we must be able to justify them. If we demand justification by reasoned argument, in the logical sense, then we are committed to the view that statements can be justified only by statements. The demand that all statements are to be logically justified (described by Fries as a 'predilection for proofs') is therefore bound to lead to an infinite regress. Now, if we wish to avoid the danger of dogmatism as well as an infinite regress, then it seems as if we could only have recourse to psychologism, i.e. the doctrine that statements can be justified not only by statements but also by perceptual experience. Faced with this trilemma – dogmatism vs. infinite regress vs. psychologism – Fries, and with him almost all epistemologists who wished to account for our empirical knowledge, opted for psychologism. In sense-experience, he taught, we have 'immediate knowledge': by this immediate knowledge, we may justify our 'mediate knowledge' – knowledge expressed in the symbolism of some language. And this mediate knowledge includes, of course, the statements of science."

11. Since its inception as a justificatory vehicle by John Rawls reflective equilibrium has been used in different philosophical contexts (1971, p. 48): "From the standpoint of moral philosophy, the best account of a person's sense of justice is not the one which fits his judgment prior to his examining any conception of justice, but rather the one which matches his judgments in reflective equilibrium. As we have seen, this state is one reached after a person has weighed various proposed conceptions and he has either revised his judgments to accord with one of them or held fast to his initial convictions (and the corresponding conception)." Reflective equilibrium is the outcome of a process by which our considered judgments of actual cases influence our moral principles, and those improved-upon principles then strengthen guidance for our response to further cases. There is an interaction

between principles and cases which are brought to cognitive closure when reflective equilibrium is attained. However, although reflective equilibrium may be the source of *coherentism* of principles and cases, it cannot provide any kind of *secure foundation*.

12. According to an anonymous reviewer "from the fact that there is no ultimate justification that may compel us to settle on the priority of one or another set of values the author concludes that we have to deal with *value pluralism*. This is the main message of Chapter 2. However, even if there were an ultimate justification of the priority of some values, we had to do with a plurality of values, and an ultimately justified ranking of our values would not imply that, for example, a top value had to be realized at any rate 'fiat iustitia et pereat mundus'. Rather we would still have to deal with a trade-off between values. We would have to choose between values, i.e. in a certain sense to evaluate them." This comment seems to strengthen the case for the necessity of coping with value pluralism.

3 NORMATIVITY

1. See Albert (1987, p. 71).
2. See Elbert (2004), Pryke (2016) and Topitsch (1958, p. 9ff.).
3. See Max Weber (1917/2004).
4. *Moral realism* is a version of realism focusing on the independence of moral entities, properties or facts and declares moral judgments to be statements that refer to facts and that can be true or false. One important facet of the debate on whether moral realism is a tenable position concerns the readiness to attribute informational content to moral judgments or rather not; one subsequently endorses *cognitivism* or *noncognitivism*. Cognitivism is in a way the epistemological and/or logical counterpart of realism, since it contends that moral judgments are moral statements that refer to moral facts and have substantial truth conditions. Noncognitivism can be viewed as the epistemological and/or logical analogue to irrealism, since it is inspired by a series of analyses, as, for example, emotivist or prescriptivist analyses, according to which moral judgments only express the emotional reactions or the prescriptions of those making them. More specifically, according to emotivism, moral judgments are not statements proper since they do not refer to facts; the sentences that we use are mere expressions

of our feelings (Ayer, 1936/1952, p. 107f; Stevenson, 1937). According to prescriptivism, on the other hand, moral judgments express prescriptions (which can be of a special universal sort (Hare, 1952) and cannot be true or false. For an overview of moral realism, see Virvidakis (1996).

5. Some prominent examples include Williams (1966/1973), Blackburn (1971/1993) and Railton (1986).
6. Just to name the most influential ones: *naturalist moral realism*, defended by philosophers who believe that moral facts and properties supervene on natural facts or properties and moral judgments are true in virtue of these facts (Boyd, 1988; Brink, 1989; Sturgeon, 1988; Copp, 2015); the *conceptual analysis* of normative statements, which typically takes the form of a biconditional, with the sentence that can be used to make a certain normative statement on the left-hand side, so that the whole biconditional amounts to a specification of a condition that is both necessary and sufficient for the truth of that normative statement (Smith, 1994; Jackson and Pettit, 1995; Jackson, 1998 and for a criticism Hatzimoysis, 2002); the *constructivist approach* of neo-Kantian theorists, which puts emphasis on moral judgments being constructions of reason and stresses that even if it were supposedly possible to grasp normative facts, we would still have to do something with them as, for example, in Korsgaard's (1996), *The Sources of Normativity* – see also her characteristic remark in Korsgaard (2003, p. 110): "If to have knowledge is to have a map of the world, then to be able to act well is to be able to decide where to go and to follow the map in going there. The ability to act is something like the ability to *use* the map, and that ability cannot be given by *another* map. (Nor can it be given by having little normative flags added to the map of nature which mark our certain spots or certain routes as good. You still have to know how to use the map before the little normative flags can be of any use to you)"; the *expressivist approach*, which suggests that normative judgments express the acceptance of systems of norms or of plans (Gibbard, 1990, 2003, 2014; Schroeder, 2008; Wodak, 2017) and in its quasi-realist version puts an emphasis on explaining why we are entitled to act as if moral judgments are truth-apt even while strictly speaking they are neither true nor false in any robust sense (Blackburn, 1993, 1988); the *quietist realist* approach, which rejects all forms of anti-realism, but

does not attempt to offer any substantive or illuminating explanation in this area of philosophy remaining a primarily negative or critical approach (McDowell, 1998); the *Platonist* approach encapsulated in the view that the intentional is normative (See, e.g., Wedgwood, 2007, p. 3: "In a way, the position that I am trying to defend can be regarded as a form of Platonism about the normative. Indeed, the doctrine that the intentional is normative can be viewed as a way of cashing out Plato's metaphor that the Form of the Good is to the understanding what the sun is to vision (*Republic*, 507b–509a). We count as sighted because we are appropriately sensitive to light, the ultimate source of which is the sun; in a similar way, we count as thinkers because we are appropriately sensitive to normative requirements, the source of which is a coherent system of eternal and necessary truths about what we ought to think or do or feel."); *Non-Naturalist Cognitivism* according to which some claims are irreducibly normative, and such claims, when they are true, do not state natural facts, but irreducibly normative facts (Parfit, 2011, Part Six). A Non-Naturalist version of value realism is contained in Thomas Nagel, 2012, chapter 5, who discusses and rejects Sharon Street's 2006 arguments that moral realism is incompatible with Darwinism.

7. For a competent review with further references, see Kristin Andrews (2016).
8. Contemporary research seems to vindicate David Hume's position in his *Treatise of Human Nature* (1748/1968, p. 176): "Next to the ridicule of denying an evident truth, is that of taking much pains to defend it; and no truth appears to me more evident than that beasts are endowed with thought and reason as well as man. The arguments are in this case so obvious, that they never escape the most stupid and ignorant."
9. See, e.g., Penn and Povinelli (2007) and Mercier and Sperber (2017, ch. 6).
10. See, e.g., Boyd and Richerson (1985) and (2009), Sterelny (2012) and Tomasello (2016).
11. See Damasio (1994), especially chapters 7 and 8.
12. See Damasio (1999, p. 284): "The inescapable and remarkable fact about these three phenomena – emotion, feeling, consciousness – is their body relatedness. We begin with an organism made up of body proper and brain, equipped with certain forms of brain response to certain stimuli and with the ability to represent the internal states caused

by reacting to stimuli and engaging repertoires of preset response. As the representations of the body grow in complexity and coordination, they come to constitute an integrated representation of the organism, a proto-self. Once that happens, it becomes possible to engender representations of the proto-self as it is affected by interactions with a given environment. It is only then that consciousness begins, and only thereafter that an organism that is responding beautifully to its environment begins to discover that *it* is responding beautifully to its environment. But all of these processes – emotion, feeling, and consciousness – depend for their execution on representations of the organism. Their shared essence is the body." For a meta-analytic summary of the neuroimaging literature on human emotion supporting the thesis that discrete emotion categories are constructed of more general brain networks not specific to those categories, see Lindquist, Wager, Kober, Bliss-Moreau and Barrett (2012).

13. Gigerenzer, Todd, and the ABC Group (1999). See also Gigerenzer, Hertwig, and Pachur (2011).
14. For an accessible review of empirical findings, see Gigerenzer (2007).
15. See Gigerenzer and Selten (2001).
16. For a useful review, see Mark Blaug (1997, chapter 8). For details of the process of transition from the classical political economy to marginal utility theory, see Mantzavinos (2016, chapter 6).
17. See Carl Menger (1871/1968, p. 86) and the English translation (1871/1976, p. 120).
18. See the impressive review of Sanjit Dhami (2017).
19. See, e.g., the works of the Nobel Laureates in Economics Douglass C. North (1990) and (2005), Oliver Williamson (1985) and (1996), and Elinor Ostrom (1990) and (2005).
20. See Elster (1989a) and (2015, chapter 21).
21. See Coleman (1990a) and (1990b). See also Ellickson (1991), McAdams (1997), Nee and Ingram (1998).
22. See Opp (1982) and Bandura (1986).
23. See Axelrod (1986) who characteristically states on p. 1097: "A norm exists in a given social setting to the extent that individuals usually act in a certain way and are often punished when seen not to be acting this way."
24. See Newell and Simon (1972).

25. See Holland et al. (1986, p. 10ff.).
26. See Ryle (1949).
27. See the remark of Ryle (1949, p. 41): "We learn how by practice, schooled indeed by criticism and example, but often quite unaided by any lessons in the theory." See also Polanyi (1958, p. 62): "The unspecifiability of the process by which we thus feel our way forward accounts for the possession by humanity of an immense mental domain, not only of theoretical knowledge but of manners, laws and of many different acts which man knows how to use, comply with, enjoy or live by, without specifiably knowing their contents." Donald (1991, ch. 6) argues for a distinct evolutionary episode in the transition from ape to human that he calls 'mimetic culture' and which still exists in modern culture. For an excellent review of the modern discussions on the issue of tacit and implicit knowledge from the perspective of modern epistemology, see Davies (2015).
28. For an overview, see Anderson (2010, chapter 9). Stanley and Williamson (2001) have contested the view that "knowing that" and "knowing how" are two distinct *kinds* of states of knowledge, and they defend the view that all "knowing how" is "knowing that." This heretic position has been heavily criticised, for example, by Schiffer (2002), John Koethe (2002) and Rosenfeld (2004). In their original contribution, Stanley and Williamson did not consider any kind of experimental evidence produced by cognitive science. Stanley (2011, chapter 7) discusses the findings of cognitive science but reinterprets the classic distinction between declarative and procedural knowledge in such a way as to fit his own view (p. 152): "The debate about declarative and procedural knowledge does not concern the existence of states of knowledge that lack truth-evaluable consent. It is rather a dispute about how to *implement knowledge*, i.e. how best to derive propositional knowledge states, procedurally or declaratively." Such an interpretation of the masses of empirical evidence available over the last decades in cognitive science is hardly convincing. For an emphatic criticism of the approach of Stanley and Williamson and more specifically of their choice to thoroughly neglect dozens of very diverse empirical studies, see Devitt (2011). Wallis (2009) and Adams (2009) also discuss a series of empirical studies from neuroscience and experimental psychology and convincingly show that Stanley and Williamson are in error.

29. I would like to stress that the term 'utility' is used here very broadly and for lack of a better term to capture the intuition which has been prevalent throughout Western philosophy, originating in Plato's *Symposium* 204e–205a:

    "'What is the love of the lover of good things?'
    'That they may be his,' I replied.
    'And what will he have who gets good things?'
    'I can make more shift to answer this,' I said; 'he will be happy.'
    'Yes,' she said, 'the happy are happy by acquisition of good things, and we have no more need to ask for what end a man wishes to be happy, when such is his wish: the answer seems to be ultimate.'
    'Quite true,' I said."

    and in Aristoteles's *Nicomachean Ethics* (1097a33): "accordingly a thing chosen always as an end and never as a means we call absolutely final. Now happiness above all else appears to be absolutely final in this sense" (The Greek text is: "καὶ ἁπλῶς δὴ τέλειον τὸ καθ᾽ αὑτὸ αἱρετὸν ἀεὶ καὶ μηδέποτε δι᾽ ἄλλο. τοιοῦτον δ᾽ ἡ εὐδαιμονία μάλιστ᾽ εἶναι δοκεῖ").
30. See Holland et al. (1986, p. 18ff.).
31. See Nersessian (2008, p. 93) who describes a mental model as a "structural, behavioral, or functional analog representation of a real-world or imaginary situation, event or process. It is analog in that it preserves constraints inherent in what is represented." See also Johnson-Laird (1983) and (2006) and the list of hundreds of articles on mental models included at the website: www.mentalmodelsblog.wordpress.com.
32. This view has been originally propagated by Konrad Lorenz (1941), Donald T. Campbell (1965) and Karl R. Popper (1972).
33. See Schrödinger (1958, p. 7): "The fact is only this, that new situations and new responses they prompt are kept in light of consciousness and well-practiced ones are no longer so (kept)."
34. John Bargh and Tanya Chartrand have influentially advocated the thesis that the process of automation is itself automatic (1999, p. 469): "But what we find most intriguing, in considering how mental processes recede from consciousness over time with repeated use, is that the process of automation is itself automatic. The necessary and sufficient ingredients for automation are frequency and consistency of use of the same-set component mental processes under the same circumstances –

regardless of whether the frequency and consistency occur because of a desire to attain a skill, or whether they occur just because we have tended in the past to make the same choices or to do the same thing or to react emotionally or evaluatively in the same way each time. These processes also become automated, but because we did not start out intending to make them that way, we are not aware that they have been and so, when that process operates automatically in that situation, we aren't aware of it."

35. See Nisbett and Ross (1980, p. 10ff.).
36. See Gigerenzer, Todd, and the ABC Group (1999) and Gigerenzer, Hertwig, and Pachur (2011).
37. See Wimsatt (2007, p. 10): "Heuristic principles are most fundamentally neither axioms nor algorithms, though they are often treated as such. As a group, they have distinct and interesting properties. Most importantly, they are re-tuned, re-modulated, re-contextualized, and often newly reconnected piecemeal rearrangements of existing adaptations or exaptations, and they encourage us to do likewise with whatever we construct."
38. See Holyoak and Thagard (1995) and Gentner, Holyoak and Kokinov (2001).
39. I cannot go into details here and the literature on choice in the behavioural and social sciences is huge. For a useful review, see Dhami (2017, Part 1).
40. G. L. S. Shackle stated it with unparalleled lucidity (1979, p. 56): "[I] have sought to avoid the word 'future'. 'Future' means something which 'is to be'. But we are here saying that there is no general evolution of human affairs which 'is to be'. The two views, of a future which pre-exists human cognisance and merely waits to be discovered, and which by greater skill and assiduity could be *discovered before its time*, and the view that the content of time-to-come is in some respects the product of present *beginning*, of origination involving something *ex nihilo*, are essentially and fundamentally in conflict."
41. See Denzau and North (1994).
42. See D'Andrade (1995) and Paul DiMaggio (1997).
43. See Ferguson (1767/1966, p. 477). See also Hume (1740/1978, p. 529f.).
44. See North, Wallis and Weingast (2009).

45. See Donald (1991, p. 344).
46. See the pioneering work of Ullmann-Margalit (1977). For a review, see Gintis (2009). See also the influential account of the emergence of social norms of Christina Bicchieri (2005) and (2016).
47. See Laudan (1977, p. 16f.).
48. See Laudan (1977, p. 16f.).

4 THE INFORMAL INSTITUTIONS OF SCIENCE

1. In the last decades, we have been witnessing the development of a research program, New Institutionalism, which has provided a series of mechanisms for the explanation of the ways that institutions shape human interaction in society, markets and politics. New Institutional Economics, for example, has become widely accepted, mainly as it has been shaped by the works of Ronal Coase (1937, 1960), Douglass C. North (1981, 1990, 1994, 2005, 2009) and Oliver Williamson (1985, 1996) who all won the Nobel memorial prize in Economic Sciences. In Sociology and Political Science, New Institutionalism has been shaped by the work of a series of authors in three different versions (Hall and Taylor, 1998), that is, historical institutionalism (Hall, 1986; Pierson, 2004; Thelen, 2004; Mahoney and Thelen, 2015), rational choice institutionalism (Riker, 1980; Alt and Crystal, 1983; Shepsle, 1986, 1989, 2006; Levi, 1988; Knight, 1992; Tsebelis, 2002, 2017) and sociological institutionalism (March, 1999; DiMaggio and Powell, 1991; Nee and Brinton, 1998; Meyer, 2010) culminating with the Nobel memorial prize in Economic Sciences awarded to the political scientist Elinor Ostrom for her interdisciplinary work on institutional analysis (1990, 2005). On the ongoing discussion on New Institutionalism in Anthropology, see Ensminger (1992, 2014), and on Classics, see Ober (2015) and Bresson (2016).
2. This solid empirical finding, that the limited cognitive faculties of the mind are used thriftily, has been presaged by Whitehead already in 1911 (p. 45f.): "It is a profoundly erroneous truism, repeated by all copy-books and by eminent people making speeches, that we should cultivate the habit of thinking of what we are doing. The precise opposite is the case. Civilization advances by extending the number of operations which we can perform without thinking about them. Operations of thought are

like cavalry charges in a battle – they are strictly limited in number, they require fresh horses, and must only be made at decisive moments."

3. For such an analysis, see Mantzavinos (2001, Part II).
4. Rowbottom's (2014) paper which criticizes van Fraassen's views as well as the views of other philosophers with respect to the 'aim of science' has the illuminating title "Aimless Science." See also the discussion in Elliott and McKaughan (2014).
5. Werner Jaeger's monumental work *Paideia* provides a panorama of the still quite limited, but nevertheless diverse educational activities in Ancient Greece (Jaeger, 1936–1947). A characteristic fixation of the diversity of cognitive praxis is contained in the standard textbook of the Middle Ages, Matianus Capella's *On the Marriage of Philology and Mercury*, written most probably at the beginning of the fifth century ACE. It describes the celestial union between the bridegroom Mercury who stands for articulateness and persuasiveness which are embodied in the three disciplines, *grammar*, *logic* and *rhetoric* and the bride Philology who stands for the affection of inquiring about the world, taking shape in the four disciplines of *arithmetic*, *geometry*, *music* and *astronomy*. The wedding is thus a metaphor of the productive synthesis of the seven liberal arts, consisting of three "humanities" and four "sciences" – all important modes of successfully coming to grips with epistemic problems.
6. Some disagree. For a prominent voice against this view, see Shapin (1996), whose work opens with the characteristic phrase: "There was no such thing as the Scientific Revolution, and this is a book about it."
7. For a very useful, accessible discussion of the question whether the Scientific Revolution was a revolution proper, see the epilogue of Cohen (2015). According to his view, the Scientific Revolution should be characterized as such, though it should not be seen as a single homogeneous event, but rather as six closely interlinked episodes of revolutionary transformation (p. 261). This view is meant to offer an answer to the argument that a full century is a very long time for a revolution. In any case, Cohen (2015, p. 257) remarks that: "[t]he contrast between 1600 and 1700 is also vast in terms of substantive content. The Earth turns. Our blood circulates. The air that we breathe has weight. A void space can be created. Objects that are attracted by the Earth fall to the ground with uniform acceleration. White light is composed of all the colours of the rainbow. Unequal cross-sections

of a river discharge equal quantities of water in equal periods of time. Human sperm consists of millions of tiny creatures. For the cycloidal trajectory of a pendulum the duration of an oscillation is independent of the deflection. The natural tones of a trumpet are linked up with multiple vibrations. The path of a bullet is a parabola. The digestive system neutralises the caustic effects of acid. Planetary orbits obey the area law. Electric repulsion exists. The list can be multiplied a hundredfold."

8. There is, of course, disagreement in the literature about this, sometimes quite extreme. Trout (2016, p. 119) writes, for example: "The truth is, given the huge lurches of theoretical progress documented in the historical record, the only way to explain the spectacular success of modern science is to assign a major role to contingent or serendipitous discovery and advance. Once we give due weight to contingency in scientific progress, we can for the first time give a more accurate account of the history of science."
9. In terms of standard game theory, as soon as one equilibrium has emerged no one has an incentive to deviate from this equilibrium.
10. See Locke (1690, p. 353).
11. The publication of the first scientific journals meant to serve as an official repository of research should be especially mentioned in this context. The first was the *Journal des Sçavans* published in 1665 followed only two months later by the *Philosophical Transactions of the Royal Society for the Advancement of Natural Knowledge.*
12. The distinction between context of discovery and context of justification has been introduced by Reichenbach (1938, p. 6f.) and extensively used by Popper (1934/2003); Arabatzis (2006) has contested it. Koertge (2000) has complemented it with the context of application. See also Henschen (2021).
13. Kit Fine has remarked in personal communication that it is an open question whether or not other kinds of representational entities might all be *identified* with propositions. A graph, for example, might be identified with the proposition that certain arguments are suitably correlated with certain values. He acknowledged, however, that other examples might be harder to deal with. It is indeed the case that, in principle, the set of rules of a diagrammatic representation of a phenomenon can be translated to a set of rules of a linguistic representation of the same phenomenon. However, a serially formatted

representation would not be concise (Teller, 2008, p. 435), since in order to describe all the different kinds of relations, a series of statements would be required and a long conjunction thereof. More importantly, the visible form of such a linguistic representation bears no relation to the structure of the phenomenon it is supposed to depict. And although some linguistic and diagrammatic representations are informationally equivalent, this does not imply computational equivalence nor does it invalidate the fact that diagrams can be more efficient platforms for drawing inferences than informationally equivalent linguistic representations (Larkin and Simon, 1987). And *pictorial representations* are not even in principle translatable into linguistic representations.

14. This is, I think, a very important distinction that the authors who insist in speaking about "technoscience" tend to oversee.
15. This is the position of Papineau (1999, p. 18) in his excellent paper on *Normativity and Judgment* in which he is "taking an implicit stance on the analysis of truth-conditional content. In particular, I am taking a stand against those approaches to content which place normativity *inside* the analysis of content, in the sense that they presuppose sui generis norms governing judgment in explaining truth-conditional content, and hence truth. Let us call theories of this kind 'non-naturalist' theories of content. Neo-verficationist and Dummettian approaches to content are of this kind, since they take content to depend on the conditions in which you are *entitled to assert* a claim. So also are Davidsonian theories of content, which take content to depend inter alia on facts about when it is reasonable to form a belief."
16. See the archetypal expression of this conception in Aristotle's Metaphysics (1011b25): "τὸ μὲν γὰρ λέγειν τὸ ὂν μὴ εἶναι ἢ τὸ μὴ ὂν εἶναι ψεῦδος, τὸ δὲ τὸ ὂν εἶναι καὶ τὸ μὴ ὂν μὴ εἶναι ἀληθές" (To say that what is not, or that what is not is, is false; but to say that what is, and what is not is not, is true).
17. For a panorama of original work on the different conceptions of truth, see Blackburn and Simmons (1999), and for a collection of contemporary work, see Glanzberg (2018). See also the overview by Raatikainen (2022).
18. The discussion that is more relevant for our context is the one on epistemic contextualism, but it does not do justice to the complexity of the situation in science. For a review with further references, see Rysiew, Patrick (2021).

19. The proposal of using speech act theory in the tradition of Austin (1962) focussing specifically on assertion is somehow more realistic, but it does not do justice to the specific institutional rules guiding scientific practice as an evidential exercise. Franco (2017) distinguishing between the cognitive attitude of accepting a hypothesis and the illocutionary force of an assertion focusses on the communicative aspects of asserting the truth or the empirical accuracy of a hypothesis. He also stresses the perlocutionary effects of an assertion on the audience. Besides the general problem of speech act theory having to do with the conceptualization of speech as an action in the first place, a conceptualization that makes it impossible to explain the ubiquitous phenomenon of human beings *saying* one thing and *doing* another (and *acting* strategically when lying), the main defect of this approach is that it conceptualizes the situation, as if the process of truth-finding were identical with the process of communicating the truths, once found.
20. Dorato (2004, p. 72) defends a similar position stressing the intersubjective validity of the test (which I will discuss later in the text): "Note, however, that the methodological problems of setting the parameters for the significance of the test do not jeopordize at all the *intersubjective validity* of the empirical claim made by the statistical hypothesis. In fact, regardless of the standard (strict or lax) one chooses according to one's preferred nonepistemic values, both parties – the overregulators and the underregulators – agree that in both cases our statistical hypotheses carry a measurable margin of error. Namely, they agree that *by choosing a stricter standard dioxin will appear to be less dangerous than it is, and by choosing a laxer standard dioxin will appear to be more dangerous.*"
21. See the excellent discussion provided by Arabatzis and Kindi (2013).
22. Resnik (1998, p. 53) includes the commitment to sincerity or honesty in his principles of the ethics of science. McMullin in his classic article on "Values in Science" stresses that even the positivists accepted the importance of moral rules for the scientific endeavour. The only difference here is that I explicitly include them as a constitutive component of the Scientific Method. See McMullin (1983/2012, p. 691): "Nor am I concerned here with *ethical* values. Weber and the positivists of the last century and this one recognized that the work of science makes ethical demands on its practitioners, demands of honesty, openness, integrity. Science is a communal work. It cannot succeed

unless results are honestly reported, unless every reasonable precaution be taken to avoid experimental error, unless evidence running counter to one's own view is fairly handled, and so on. These are severe demands, and scientists do not always live up to them. Outright fraud, as we have been made uncomfortably aware in recent years, does occur. But so far we can tell, it is rare and does not threaten the integrity of the research enterprise generally. In any event, there never has been any disagreement about the value-ladenness of science where moral values of this kind are concerned."

23. On epistemic trust in science, see Wilholt (2013).
24. Glymour (1980, p. 12f.), himself a proponent of the bootstrap view of confirmation, provides a concise summary of the main strategies offered: "We find phenomenalist and nonphenomenalist alike using the same methods, and the successes and failures of the methods were largely independent of the epistemological theories in which they may have been embedded. It is worthwhile listing at least the principal kinds of strategies that seem to have been employed:

    1. *Elimination of theory*: Sentences whose vocabulary wholly or partly transcends that permitted for description of the ultimate evidence are to be eliminated in favour of statements entirely within the language of evidence. The elimination is to be accomplished by somehow reducing statements of the broader language to statements in the narrower one.
    2. *The deductive method*: At one time, anyway, the most popular version of the evidential connection took it to depend on a deductive connection going on the opposite direction. Evidence confirms a theory if the evidence can be deduced, in an appropriate way, from the theory. It may, for all I know, still be the most popular version.
    3. *The bootstrap method*: In essays whose point has been lost with the passage of philosophical time, both Carnap and Reichenbach proposed a strategy that was developed so explicitly nowhere else (although it was outlined by Weyl and others): theoretical hypotheses are confirmed by deducing them or positive instances of them from observation statements. Because the vocabulary of the theory extends beyond the vocabulary of observation statements, the deduction requires as premises hypotheses containing both theoretical and observational predicates. Carnap and Reichenbach regarded these

auxiliary hypotheses as an epistemically privileged – analytic – part of the theory.
4. *Probabilistic strategies*: Typically, these theories suppose that there is a conditional probability function on pairs of sentences, and in particular that there is a conditional probability for any hypothesis given any evidence statement. Confirmation relations are analyzed in terms of probability relations."

25. See Psillos (2015, p. 364) for a discussion of this point.
26. Norton (2008, p. 18) calls this the "gap argument." So does Matthew Brown (2013, p. 831): "The many forms of underdetermination arguments have in common the idea that some form of gap exists between theory and observation. Feminists, pragmatists, and others have sought to fill the gap with social values or to argue that doing so does not violate rational prescriptions on scientific inference. Call this the *gap argument* for value-laden science."
27. See the beginning of Norton's (1993, p. 1) paper dealing with the case of quantum discontinuity, 1900–1915: "There is a serious contradiction between a thesis increasingly popular amongst philosophers of science and the proclamations of scientists themselves. The underdetermination thesis asserts that a scientific theory cannot be fully determined by all possible observational data. Scientists, however, are not so pessimistic about the power of observational data to guide theory selection. The history of science is full of cases in which they urge that the weight of observational evidence forces acceptance of a definite theory and no other. Thus our science text books teach us to accept the approximate sphericity of the earth, the heliocentric layout of planetary orbits, the oxygen theory of combustion, and a host of other theoretical claims simply because the evidence admits no alternatives. The case for the underdetermination thesis depends in large measure on an impoverished picture of the ways in which evidence can bear on theory."
28. As Norton (2008, p. 18f.) correctly remarks: "The underdetermination thesis has long been a truism in science studies, accepted and asserted with the same freedom that philosophers now routinely remark on the impotence of logic to provide us with a finite axiomatization of arithmetic. Yet the underdetermination thesis enjoys no such status in philosophy of science, where it is hotly debated." And continues (p. 21f.): "[…] I should mention one basis for the thesis that withstands

little scrutiny: the thesis has become such a commonplace that it has entered that elusive body of knowledge labelled 'what everyone knows'. That is, it is justified by the fallacy, *argumentum ad populum*."

29. See the review in the Excursus, and for the current discussion with further literature, see footnote 4. ChoGlueck (2018) claims that the argument from inductive risk, which he calls the error argument, is nested within the underdetermination thesis, which he calls the gap argument – according to the slogan "the error is in the gap."
30. To quote again from Norton's paper (1993, p. 31): "The underdetermination thesis tells us that theory remains underdetermined by any body of evidence, no matter how large, rich, and diverse. If it were true, the theoretician seeking to build a theory on a body of evidence might reasonably expect to be faced with a plethora of theories, all of which do justice to the evidence. Yet this is not the common experience. When the available evidence is substantial, theoreticians consider themselves lucky to find *any* theory that does justice to the evidence and, if they do find one, the construction of competitors with any long-term viability proves well-nigh impossible. Perhaps this phenomenon is due to prejudice, social conditioning, stupidity, deference to dictatorial authority, or a host of other distractions that are traditionally deplored non-scientific. What this study shows, however, is that, at least in the instance of quantum discontinuity, whichever of these forces were in operation, the weight of evidence was sufficient to force the unique determination of a particular result, no matter how unpalatable the result might be to the community of physicists. [...] I do not believe that the case is exceptional. Rather, it merely illustrates a commonplace of the lore of science, the power of a sufficient body of evidence to determine a unique theory."
31. See the excellent discussion by Deaton and Cartwright (2018) about what randomized controlled experiments can convey and about their scope and limitations. For a good review of evidence in epidemiology, see Plutynski (2018, p. 131ff.).
32. It is impossible to provide a full-blown theory of evidence here. Reiss in developing a pragmatist theory of evidence, as he calls it, delineates the desideratum (2015, p. 343): "[T]o collect facts and to make up one's mind (i.e., to infer a hypothesis) are two different activities. 'Support' relates to the collection of facts; 'warrant' relates to making up one's

mind. 'Evidence', unfortunately, conflates the two. A good theory of evidence should explicate both support and warrant. We need, on the one hand, criteria or guidelines that tell us what kinds of facts we have to collect in order to evaluate a hypothesis; we need to know what facts are relevant to the hypothesis. We need, on the other hand, criteria or guidelines that tell us how to assess the hypothesis, given the facts we 've collected in its support, or, conversely, criteria or guidelines that tell us how much support of what kind we need in order to achieve a given degree of warrant. We require criteria or guidelines that translate between knowledge of the facts relevant to the hypothesis and judgments about the hypothesis. A theory of evidence that didn't tell us about relevance would be impracticable; a theory that didn't tell us about assessment would not be useful. Here, then, is a first desideratum [...]: the theory should be a theory of both support and warrant."

33. I want to clarify that my claim that truth is an *epistemic* property of a representational entity, should not be confused with the endorsement of any kind of *epistemic theory of truth*. Such theories of truth contend that truth is constitutively connected to epistemic concepts such as warrant, justification or verification – such epistemic concepts are taken as more primary, so that they can be used in order to define truth. This is not the position taken in the text.
34. See the useful discussion about truthlikeness in Niiniluoto (2023, p. 32ff.).
35. Catherine Elgin (2017) has introduced this idea.
36. For a useful discussion, see Schindler (2018, p. 7ff.).
37. A theory does not have a purpose or adequacy conditions either, as Brigandt (2015, p. 343) argues.
38. McMullin (1983/2012, p. 700) provides the example of the quantum theory of matter: "The notorious disagreement between Bohr and Einstein in regard to the acceptability of the quantum theory of matter did not bear on matters of predictive accuracy. Einstein regarded the new theory as lacking both in coherence and in consistency with the rest of physics. He also thought it failing in simplicity, the value that he tended to put first. Bohr admitted the lack of consistency with classical physics, but played down its importance. The predictive successes of the new theory obviously counted much more heavily with him than they did with Einstein. The differences between their assessments were not solely due to differences in the values they employed in

theory-appraisal. Disagreement in substantive metaphysical belief about the nature of the world also played a part. But there can be no doubt from the abundant testimony of the two physicists themselves that they had very different views as to what constituted a 'good' theory."

39. Sober (2002, p. 15) has proposed to view simplicity of theories as the minimization of something: "The minimization might involve a semantic feature of what a set of sentences says or a syntactic feature of the sentences themselves. An example of the first, semantic, understanding of simplicity would be the idea that simpler theories postulate fewer causes, or fewer changes in the characteristics of the objects in a domain of inquiry. An example of the second, syntactic understanding of simplicity would be the idea that simpler theories take fewer symbols to express, or are expressible in terms of a smaller vocabulary."
40. Dorato (2004, p. 54) puts it succinctly: "[T]he causal origin of any argument – the reason why anybody would want it to be valid – has nothing to do with its validity."
41. Wilholt (2022, p. 88) expresses the fear that "[w]ithout a substantial ideal of objectivity (as an *epistemic* ideal) researchers could not be criticized for allowing idiosyncratic value influences (such as the profit interests of a particular branch of industry) to have an impact on their research, but only for permitting the 'wrong' kinds of values (i.e., others than those preferred by the critic)." But what forbids anyone to invoke scientific objectivity in the procedural sense proposed here to criticize researchers both on *moral* grounds and on *grounds having to do with the improper use of scientific techniques*?
42. See the remark of Hoyningen-Huene (2023, p. 14f): "Clearly, the notion of objectivity is somehow related to the notion of truth in the correspondence sense. In both cases, some sort of reference to 'mind-independent' or 'subject-independent' facts or objects is implied [...] However, an obvious difference concerns the possibility of comparative use, which 'objective' smoothly allows ('a is more objective than b') and which sounds somewhat awkward for 'true' ('a is truer than b'). The reason is that 'objective' contains the additional meaning component of fairness and balance, that is, the rejection of an unbalanced ('subjective') selection of features of the represented object, and fairness and balance come in degrees. The additional meaning component of 'objective' becomes especially obvious when we consider an extreme case. Imagine

a report of a demonstration of 10,000 people in which 100 active rioters participate. Suppose the report hardly mentions the peaceful participants but extensively describes the actions of the rioters such that the impression arises that the demonstration was fairly violent. Although every single sentence in the report may be true, the report will not count as objective because the selection of reported features of the demonstration is unbalanced and one-sided."

43. Holman and Wilholt (2022, p. 214) in the course of introducing what they call 'the new demarcation problem' lucidly make the claim that the traditional desirability of the value-free ideal served three purposes: "*Veracity*: Scientists should pursue the discovery of knowledge. *Universality*: Scientists should produce results useable by anyone for purposes not anticipated by the researcher. *Authority*: Scientists should propose a trustworthy body of knowledge that has broadly recognized social legitimacy." I believe that my approach satisfies all three desiderata. For a discussion of the 'new demarcation problem', see Resnik and Elliott (2023).
44. "Personally, I am deeply interested in science and value it enormously. When you need something, do appeal directly to me."
Lenin to Sergei Oldenburg, January 27, 1921
Quote from the book of Krementsov (1997, p. 13) on *Stalinist Science*.
45. Daston (1992) provides a wonderful account of how 'aperspectical objectivity' came to become the ideal of scientific objectivity only in the nineteenth-century science and stresses that the essence of aperspectival objectivity is communicability (p. 69). See also her brilliant remark (p. 609): "Aperspectival objectivity was the ethos of the interchangeable and therefore featureless observer – unmarked by nationality, by sensory dullness or acuity, by training or tradition; by quirky apparatus, by colourful writing style, or by any other idiosyncracy that might interfere with the communication, comparison and accumulation or results. Scientists paid homage to this ideal by contrasting the individualism of the artist with the self-effacing cooperation of scientists, who no longer came in the singular – 'l'art c'est moi, la science, c'est nous', in Claude Bernard's epigram." On the history of objectivity, see the landmark work by Daston and Galison (2007).
46. See Nagel (1986, p. 5) who also admits of degrees of objectivity: "A view or form of thought is more objective than another if it relies less on the specifics of the individual's makeup and position in the world, or on

the character of the particular type of creature he is." See also Williams (1985, p. 139) who endorses also an absolute conception for science, one that seeks "to represent the world in a way to the maximum degree independent of our perspective and its peculiarities."

47. This characterization is, as Fine (1998, p. 9) stresses, due to Arthur Eddington (1920, p. 30) who was the leader of the British expedition of 1919 that took place in order to test the first dramatic prediction of Einstein's theory, the bending of light rays around the sun: "The first step in throwing our knowledge into a common stock must be the elimination of the various individual standpoints and the reduction to some specified standard observer. The picture of the world so obtained is none the less relative. We have not eliminated the observer's share; we have only fixed it definitely. To obtain a conception of the world from the point of view of no one in particular is a much more difficult task."
48. Fine (1998, p. 18) claims that there is what he calls a 'classical process-product fallacy' and that the objectivity of an outcome does not follow from the objectivity of a process. He gives the example that 'there are safe processes for producing bombs, and bombs are not safe products' – but this is, of course, not an expedient example, because it is a property of the bombs not to be safe. As Koskinen (2020, p. 1194, Fn. 2) correctly observes "[i]f a process is designed so as to avoid some specific form of 'scientific subjectivity', it is not a fallacy to conclude that the product is not distorted by that form of subjectivity."
49. See Sankey (2019, p.7): "Scientific inquiry is characterized by widespread consensus among scientists. Disagreement, where it occurs in science, is short-lived. The reason that science is characterized by consensus is precisely due to epistemic objectivity and the role played by the scientific method in ensuring such objectivity. It is because scientists employ a shared scientific method that they come to agree with each other. The shared scientific method ensures epistemic objectivity, and thereby promotes the formation of consensus among scientists.
    But what is the scientific method and how does it produce consensus? According to traditional empiricist philosophy of science, the method of science consists of two key elements: observation and inference. Observation itself is objective because human observers are endowed with a common perceptual apparatus which provides them with perceptual access to the unique, publicly accessible objective world. Inference – whether inductive or deductive – has a shared logical structure, so that

all scientists draw the same conclusions from the same empirical data. In this way, the fact that scientists all employ the same scientific method is what gives rise to the widespread consensus that characterizes the sciences." For an overview of the discussions on scientific objectivity, see Reiss and Sprenger (2020) and John (2021). Douglas (2004) argues for the irreducible complexity of objectivity articulating eight distinct senses of objectivity. Koskinen (2020) on the contrary stresses the commonalities of the different uses of the concept of objectivity.

50. Arthur Fine (1998, p. 13) stresses that there are not only two extreme alternatives in the form of either perfectly general principles (or standards) or of no principles (or standards) at all.
51. As Arthur Fine (1998, p. 16) succinctly remarks: "However there is no magic method that is reliable all around. Science proceeds on the basis of trial and error and what happens in most laboratories and in most centers of calculation on most days in most years is the methodical, procedurally objective production of errors."
52. Larry Laudan's (1987) normative naturalism is according to these lines, without however a commitment to truth.
53. Anke Bueter (2015) draws attention to what she calls 'blind spots' in the context of discovery and contends that the theory assessment in the context of justification is not independent from the values in the context of discovery. To use the term 'blind spots' is surely an understatement of the situation; on the other hand, if the view of scientific objectivity proposed in the text is adopted, then the epistemic products that are scientifically warranted become in no way less objective, because of the existence of such blind spots. On this issue, see also Elliott and McKaughan (2009).
54. Kitcher (2001, p. 79f.) contends that ascriptions of significance are possible and indeed required, because science is not after truth, but after significant truth.
55. Ian Hacking (1983, p. 263) in his *Representing and Intervening* provides the characteristic example of the case that scientists make use and therefore de facto do not doubt the objective existence of electrons when they can successfully use them to produce images of other entities with an electron scanning microscope.
56. See the discussion in the Excursus.
57. Carrier (2022, p. 17) correctly stresses just before the passage cited in the text: "[T]he trouble with giving up with value-freedom is that committing

science to certain nonepistemic goals is liable to produce a politicized and biased science that would cease to be the ecumenical source of knowledge on which all parties at strife can rely. This would further fuel public suspicion that experts are hired guns and can be rented to fight for a political cause." Even in economics, a much-contested discipline, there is an established practice distinguishing between economic theory and economic policy, the latter being specialized in formulating policy packages without committing to the nonepistemic objectives.

58. See also the broad discussion by Carrier (2017).

5 CORE SCIENTIFIC ACTIVITIES

1. The Cartesian dualism between *res cogitans* and *res extensa* was still implicit at the background of the discussion in the nineteenth century when the question was debated whether human actions were to be viewed as physical phenomena or not and how they should be treated. Naturalists since Mill (1843/1974, Book VI) have contended that actions have to be viewed as phenomena on a continuum with other phenomena in nature and that they should be studied accordingly. Issues of interpretation hardly emerge, if one adopts such a view. Interpretivists such as Dilthey (1883/1990, 1924/1990, 1927/1992), on the contrary, have argued forcefully that human actions cannot be viewed as natural phenomena since their meaningfulness makes them categorically distinct. If one adopts the interpretivist view, then issues of interpretation arise in the space of the mental. A distinct methodology for all disciplines which deal with meaningful material appears mandatory and a methodological dualism is defended.
2. The age-old "Verstehen vs. Erklären" debate in the German-speaking world was largely about the following question: whether there is a distinct method for the apprehension of meaningful material, employable in the social sciences and the humanities (*Geisteswissenschaften; Kulturwissenschaften*), which deal with such material, that is, *Verstehen* (understanding), or whether the general method employed in the natural sciences is successfully employable in the social sciences and humanities as well, that is, *Erklären* (explanation). Methodological dualists famously pleaded for the autonomy of the social sciences and humanities which must follow the method of *Verstehen*. The neo-Kantian philosophers Wilhelm Windelband and Heinrich Rickert focused on the methods of concept

formation and judgment in the different groups of sciences, the *Kulturwissenschaften* and the natural sciences. For Windelband (1894), the logic of *Kulturwissenschaften* is characterised by an *idiographic interest* in singular judgments about the past opposed to the natural sciences' *nomothetic interest* in formulating laws. For Rickert (1929), the *Kulturwissenschaften* are characterised by an individualising form of concept formation, which solved the problem of how the general concepts essential to any scientific representation could capture an individual object, without simply subsuming it under a general law in the fashion of natural concept formation. On this debate, see Lanier Anderson (2003) and Bouterse and Karstens (2015).
In the UK, the famous Rede Lecture of Snow in 1959 has conceptualised the debate more broadly in terms of his diagnosis about the existence of 'two cultures'; see Snow (1959/1993). Lastly, on the famous science wars, see the classic, but polemical, *Impostures Intellectuelles* by Sokal and Bricmont (1997), and for a more balanced perspective, see Jardine and Frasca-Spada (1997).

3. Nehamas (1985) was according to my knowledge the first to plead that even science and literary theory share common methodological grounds. He concluded his paper as follows (p. 86): "The philosophy of science used to be one of the least historically conscious branches of philosophy. Literary theory came about in large measure as a reaction to the excesses of what used to be known as literary history. It is as ironic as it is exhilarating to see that the philosophy of science now suggests that its newly found historical dimension be transferred to literary theory, while literary theory offers to the philosophy of science a broader and vastly richer notion of explanation and interpretation than had ever seemed before possible."
4. The honorific title 'science' is, however, used to characterise the work of scholars in the humanities in the German-speaking world where the term *Geisteswissenschaften* is common as well as in the French-speaking world where the term *Sciences Humaines* is common.
5. For extensive treatments of the notion of explanatory game and explanatory pluralism, see my essay *Explanatory Games* (2013) and my book *Explanatory Pluralism* (2016).
6. Bechtel and Abrahamsen (2005, p. 425) also stress the role of representation in explanation within the premises of the mechanistic account: "Thus, since explanation is itself an epistemic activity, what

figures in it are not the mechanisms in the world, but representations of them. These representations may be internal mental representations, but they may also take the form of representations external to the cognitive agent – diagrams, linguistic descriptions, mathematical equations, physical models, and so on." Bokulich (2018) makes a more general point endorsing what she calls 'the eikonic conception of scientific explanation', which is independent on any specific *account* of explanation. According to this conception, explanations are the product of epistemic activity involving representations of the phenomena to be explained, and she stresses that "[w]hat scientists explain in the first instance is not the phenomenon-in-the world itself but the phenomenon as represented" (p. 801). Such a conception is juxtaposed to the ontic conception of explanation as originally and influentially proposed by Salmon (1984, p. 17ff.) and more recently vigorously defended by Craver, who writes (2014, p. 40): "Conceived ontically, however, the term explanation refers to an objective portion of the causal structure of the world, to the set of factors that produce, underlie, or are otherwise responsible for a phenomenon. Ontic explanations are not texts; they are full-bodied things. They are not true or false. They are not more or less abstract. They are not more or less complete. They consist in all and only the relevant features of the mechanism in question. There is no question of ontic explanations being 'right' or 'wrong,' or 'good' or 'bad.' They just are." For a criticism of the ontic conception, see Wright (2015).

7. On the validity of resemblance theory in virtual representations, see Greenberg (2013), who argues that accurate depiction cannot be wholly characterised in terms of similarity or in terms of difference, but rather with the help of the broader notion of *transformation* that incorporates both. On the role of idealizations, an important issue that I cannot address here, see Weisberg (2007) and (2013) and the interesting discussion in Potochnik (2017, p. 41ff.).
8. I am deliberately using the term 'nexus of meaning' as a *terminus technicus* in order to distinguish my approach from other approaches to meaning.
9. In fact, none of the principles proposed in this discussion is new. As early as 1654, Johannes Clauberg has worked out in admirable detail principles of "*in bonam partem interpretari*" in Chapter XIII of the third part of his *Logica, Vetus & Nova*, the principle of charity – "*benignitas*" – being the most important one (Clauberg 1654). And in

1757, Georg Friedrich Meier proposed the principle of hermeneutic equity as the most general principle of all interpretative rules of a *hermeneutica universalis* (Meier 1757/1996: §39): "Hermeneutic equity (*aequitas hermeneutica*) is the tendency of the interpreter to hold that meaning for hermeneutically true that best comports with the flawlessness of the originator of the sign, *until the opposite is shown.*" It is important to stress that the principle of hermeneutic equity is explicitly formulated as a presumption: a rule that can fail to stand up to evidence.

10. An anonymous reviewer has questioned "whether simplicity still qualifies as a marker of truth for the sciences of complexity (e.g. climatology), in which tiny perturbations can cause mega-effects." I agree with this comment – it is surely the case that not *all* markers of truth can be used in the evaluation of *all* theoretical constructs.
11. An evaluation can take different forms, of course, as, for example, in the work of Kuhn (1977) with respect to theory choice, in Laudan's (1984, p. 62ff.) on the development of the reticulated model of scientific rationality and in Parker's (2020) recent adequacy-for-purpose view.
12. For a discussion, see Sen (2006) and Mantzavinos (2021a). On the logic of comparatives, see Dorr et al. (2022).
13. It is impossible to offer here a full-blown discussion of the issue. For a very useful taxonomy, see Nickles (2021). I would only like to note that most contemporary discussions oscillate between two extreme views, that is, the view that scientific rationality is encapsulated in the application of the scientific method, which yields scientific knowledge, and the view that questions the very possibility of scientific endeavours to produce reliable knowledge of natural, biological and social phenomena. Some still defend the possibility of an essentially formal rule of scientific rationality – the different versions of Bayesianism provide a good example of this position (e.g., Howson and Urbach, 2006). A series of so-called postmodern theorists, on the other hand, have radicalised the criticisms of Kuhn, Feyerabend and the strong programme in the sociology of science, essentially denying the very rationality of the scientific enterprise. There is also a series of "middle ground positions." Kitcher (1993, 2001, 2011) and Longino (1990, 2002, 2013) acknowledge the social character of science but at the same time honour the rationality of the scientific process by highlighting the possibility of acceptance or

rejection of theoretical constructs on the basis of evidence allowing scientific objectivity – with a human face – to prevail. Friedman (2001, 2002, 2011) argues for a relativised and historicised version of the original Kantian conception of scientific a priori principles and examines the way in which these principles change and develop across revolutionary paradigm shifts.

14. See the German translation in Svoronos (1908).
15. See translation by D. J. Zeyl, in Cooper (1997, p. 1244).
16. A brilliant discussion on the role of objectivity in the humanities, especially history, is provided by Daston (2014, pp. 33, 36): "[Leopold von] Ranke, whose legendary seminar was the cradle of all these 'objective measures and controls' among the historians, renounced any intention of writing vivid, edifying history: that is the context of his famous declaration that all he wanted to do was 'bloß zeigen, wie es eigentlich gewesen' – even at the price of a narrative that was 'oft hart, abgebrochen, ermüdend'. [...] The acolytes of this new and decidedly uncomfortable religion of historical objectivity were almost all formed in the new-style research seminar initiated by the reformed German universities and imitated widely throughout the learnt world by the end of the nineteenth century. It was the research seminar that in reality disciplined the disciplines. It was the prime mover behind the multiplication of specialist societies and journals. In the seminars students learnt that *Wissenschaftlichkeit* meant method, and method in turn meant the mastery of esoteric techniques through long, arduous application. Whether the technique in question was paleography learnt at the Berlin philology seminar or error analysis learned at the Königsberg physics seminar, the craft knowledge imparted by close contact of professors with students resembled nothing so much as an apprenticeship with a master. [...] *wissenschaftlich* referred almost invariably to the painstaking, abstruse techniques – those very methodical 'methods of research' – that certified a piece of work, be it an experiment or an edition, as objective."
17. Due to lack of space, I cannot provide a discussion of scientific understanding, but I may note that the recent literature on the scientific understanding that (successful) *explanations* provide is congruent with the older literature in the humanities and the social sciences of the scientific understanding that (successful) *interpretations* provide.

6 THE FORMAL INSTITUTIONS OF SCIENCE

1. I have dealt extensively with this issue in Mantzavinos (2001, chapter 8).
2. For an interesting discussion of the doctrine of *partiinost* that scientific work should be subordinated to the Party's interest, see John (2019, p. 65).
3. Competition for recognition has been paid a great deal of attention in the literature and is often conceptualised as a search for credit on the part of scientists. The original idea has been proposed by Kitcher in his paper "The Division of Cognitive Labor" (1990). On the role of the *priority rule* in this context which requires that the lion's share of credit goes to the person who discovered a phenomenon first, see Strevens (2006a). On the role of the so-called *Matthew effect* according to which, when discoveries are made nearly simultaneously, or by teams of scientists working together, most of the credit is conferred on the more famous of the discoverers, see Strevens (2006b). See also Zamora Bonilla (2002) and (2013). On the problematic that scientific outcomes have the character of public goods, see the locus classicus Nelson (1959). According to Congleton (1989, p. 185), the status-seeking behaviour can help to solve the public goods problem when those seeking status produce something that is beneficial to others. For a discussion and further references, see Zollman (2018).
4. In a paper following the general idea of a "philosophical history of science" proposed by Arabatzis (2017), Nickelsen (2022, p. 24) provides a very useful case study of the photosynthesis research communities, which "unravels how the *self-organized interplay of cooperation and competition* as well as individual and collective goals within a scientific community proved conducive to the generation of knowledge" [My emphasis, C.M.].
5. James Buchanan (1977/2001, p. 99), having the case of markets in mind, but stating a more general point about invisible hand explanations, has observed that they "may be as applicable to 'orders' that are clearly recognized to be undesirable as to those that are recognized to be desirable."

7 THE SEARCH FOR AN ADEQUATE CONSTITUTION

1. For an interesting discussion on *politeia* and *nomos*, see Lewis (2011).
2. Stressing the particular circumstances was what de Maistre must have had in mind as it is evidenced in his penetrating comment

(1794–1795/1965, p. 80): "The 1795 Constitution, like its predecessors was made for *man*. But there is no such a thing as *man* in the world. During my life, I have seen Frenchmen, Italians, Russians, and so on; thanks to Montesquieu, I even know that one may be a *Persian*; but I must say, as for *man*, I have never come across him anywhere; if he exists, he is completely unknown to me."

3. This is the title of the book by Brennan and Buchanan (1985). See also the classic work of Buchanan and Tullock (1962) and Buchanan (1975).
4. See Voigt (2020, p. 6).
5. See Hadfield and Weingast (2014).
6. For a game-theoretic account of this issue, see Weingast (1997).
7. See https://comparativeconstitutionsproject.org.
8. See https://v-dem.net.
9. See the interesting article of Elster (2012) on how to design such an assembly.
10. See Karl Popper (1934/2003, p. 33).
11. See Polanyi (1966, p. 82f.).
12. See Kitcher (2011, p. 98).
13. Amartya Sen (2006, p. 222) makes a similar point in juxtaposing what he calls the 'transcendental approach to justice' to the 'comparative approach to justice'. See also the more detailed account in Sen (2009).
14. The first who has clearly seen and articulated this problematic and introduced the term "paradox of liberal democracy" is, according to my knowledge, Aristides Hatzis (2015, 227ff.).
15. For a very useful discussion, see Wilholt (2010), who traces the variations of the argument in favour of freedom of research to the early modern defenders of the freedom of philosophizing including Campanella, Descartes, Milton and Spinoza. John Stuart Mill's famous argument for intellectual freedom as stated in his *On Liberty*, ch. 2, by appeal to the fallibility of human judgment was preceded by the German philosopher Nicolas Gundling, who had already stated it in a speech delivered at the University of Halle in 1711 and by Christian Wolff's description of the mechanism of mutual criticism in 1728. For a discussion and further references, see Wilholt (2010, p. 175).
16. There is also an international aspect to this question, but due to lack of space I cannot tackle it here. See, for example, Hans Albert (2010, p. 409).
17. It should be briefly noted that the epistemic theory of democracy (Goodin and Spiekermann, 2018), a version of democratic theory

emanating from the well-known passage from Aristotle's *Politics* (III, 11, 1281a41–1281b2) "the many, who are not as individuals excellent men, nevertheless can, when they have come together, be better than the few best people, not individually but collectively" and Condorcet's Jury Theorem (1785), does make the claim that majority voting is the appropriate method of finding out the truth. Condorcet's famous Jury Theorem states roughly two things: (a) the majority vote among a group of independent voters is itself more likely to be right than are individual voters separately and (b) as the number of such voters approaches infinity, the probability that the majority among them is correct approaches one. However, it should be clear that apart from many other problems concerning the applicability of this mathematical theorem to real-world affairs within the premises of a democratic polity on simple 'yes' or 'no' issues, its applicability to the hugely more complex epistemic problems that scientific agents are trying to solve is anything but plausible. Estlund (2008, ch. 12), an advocate of an epistemic theory of democracy, characterizes the Condorcet Jury Theorem as entirely 'irrelevant'. See also the interesting discussion in the introduction of Landemore (2012) in her co-edited volume with Jon Elster on *Collective Wisdom: Principles and Mechanisms.*

18. Schroeder (2022) in his last paper on the limits of democratizing science wants to answer the question 'when scientists should ignore the public' via discussing the views of some political philosophers. He writes that he "will ignore concerns about feasibility" (p. 2). But why invoke any philosophy at all, if something is not feasible to attain? 'The public' is not an agent, and insofar it cannot have values. In a pluralistic society organized as a polity, only *some* individuals will have *some* shared values and there is no consistent formula to aggregate inconsistent values.

    In any case, it is only a constitution that provides the trade-off between different values. As Galston (2011, p. 238) correctly observes, "every political community assumes a distinctive form and identity through its constitution. A constitution, we may say, represents an authoritative partial ordering of public values. It selects a subset of worthy values, brings them to the foreground, and subordinates others to them. These preferred values then become the benchmarks for assessing legislation, public policy, and even the condition of public culture."

19. Guaranteeing the open access to science to everybody would also solve the problems connected with the possible threats to scientific objectivity caused by extra-academic participation in the scientific knowledge production. For arguments on how, depending on the specifics of each case, so-called citizen science, participatory projects etc. may increase or threaten scientific objectivity see the interesting discussion by Koskinen (2023). Participation in the scientific process by non-experts, especially when important economic, political or other interests are at stake can go hand in hand with "lobbying, bullying, and bribery" as Wilholt (2014, p. 171) correctly observes. For a useful discussion of empirical cases of hijacking and diversity washing, see Koskinen (2022, p. 10ff).
20. See https://iranonline.com/iran/government/constitution/constitution-1/.

8 FIVE PRINCIPLES FOR A QUASI-AUTONOMOUS SCIENCE

1. See Karl Popper (1963, p. 126) "I think that the innovation which the early Greek philosophers introduced was roughly this: they began to *discuss* these matters. Instead of accepting the religious tradition uncritically, and as unalterable [...], instead of merely handing on a tradition, they challenged it, and sometimes even invented a new myth in place of the old one. [...] the Greek philosophers invented a *new tradition* – the tradition of adopting a critical attitude towards the myths, the tradition of discussing them; the tradition of not only telling a myth, but also of being challenged by the man to whom it is told."
2. See Karl Popper (1945/2002, p. 489f.) and Strevens (2020, p. 93). Some experiments in contemporary science might not be repeatable, as Theo Arabatzis has stressed to me in personal communication, like experiments in high-energy physics that require a very costly material infrastructure. They are *in principle* repeatable, however.
3. See McCauley (2000, p. 73ff.).
4. See Fernández Pinto (2020).
5. See Elster et al. (2018).

EPILOGUE

1. See Pettit (1997, p. 250ff. and 2012, p. 174).

EXCURSUS

1. The so-called *Werturteilsstreit* took place in the Verein für Socialpolitik in 1913 – see the contributions in Nau (1996). On the position of Max Weber, see the excellent discussion by Herbert Keuth (1993, part 1.) See also Albert and Topitsch (1979).
2. This term was introduced by Oppenheim (1871), a liberal thinker and politician. Most important *Kathedersozialisten* were Gustav Schmoller, Adolph Wagner among others.
3. See Max Weber (1917/1985, p. 491ff.).
4. See Albert (1963/1993) and Albert (1966, p. 203). Radnitzky (1981) calls it the "Werturteilsfreiheitsthese."
5. See Max Weber (1904/1985, p. 151).
6. See Max Weber (1904/1985, pp. 151–158) and (1917/1985, p. 500).
7. See Max Weber (1904/1985, p. 212f.): "Die *objektive* Gültigkeit alles Erfahrungswissens beruht darauf und nur darauf, daß die gegebene Wirklichkeit nach Kategorien geordnet wird, welche in einem spezifischen Sinn *subjektiv*, nämlich die *Voraussetzung* unserer Erkenntnis darstellend, und an die Voraussetzung des *Wertes* derjenigen Wahrheit gebunden sind, die das Erfahrungswissen allein uns zu geben vermag. Wem diese Wahrheit nicht wertvoll ist – und der Glaube an den Wert wissenschaftlicher Wahrheit ist Produkt bestimmter Kulturen und nichts Naturgegebenes -, dem haben wir mit den Mitteln unserer Wissenschaft nichts zu bieten."
8. See Adorno, Theodor et al. (eds.): *Der Positivismusstreit in der deutschen Soziologie*, Neuwied/Berlin: Hermann Luchterhand Verlag, 1969. On how this debate has been baptized and on the history of the publication of the volume, see Popper's essay on "Reason or Revolution?" in his *The Myth of the Framework* (1994).
9. See Habermas (1969a, p. 170ff.). Habermas referred to Popper's well-known passage from *The Open Society and Its Enemies* (1945/2002, p. 503): "The rationalist attitude is characterized by the importance it attaches to argument and experience. But neither logical argument nor experience can establish the rationalist attitude; for only those who are ready to consider argument or experience, and who have therefore adopted this attitude already, will be impressed by them. That is to say, a rationalist attitude must be first adopted if any argument or experience is to be effective, and it cannot therefore be based upon

argument or experience. (And this consideration is quite independent of the question whether or not there exist any convincing rational arguments which favour the adoption of the rationalist attitude). We have to conclude from this that no rational argument will have a rational effect on a man who does not want to adopt a rational attitude. Thus a comprehensive rationalism is untenable.
But this means that whoever adopts the rationalist attitude does so because he has adopted, consciously or unconsciously, some proposal, or decision, or belief or behaviour; an adoption which may be called 'irrational'. Whether this adoption is tentative or leads to a settled habit, we may describe it as an irrational *faith in reason*. So rationalism is necessarily far from comprehensive or self-contained." Such a position seems, according to Habermas (ibid., p. 171), to be positivistic alluding to Wittgenstein's Tractatus 6.52: "Wir fühlen, dass selbst, wenn alle *möglichen* wissenschaftlichen Fragen beantwortet sind, unsere Lebensprobleme noch gar nicht berührt sind." ("We feel that even if *all possible* scientific questions be answered, the problems of life have still not been touched at all.") [Translation by Ogden in Wittgenstein (1922, p. 187).]

10. See Habermas (1969b, p. 258): "Deshalb entgeht Popper durch Berufung auf den Korrespondenzbegriff der Wahrheit dem dialektischen Zusammenhang zwischen deskriptiven, postulatorischen und kritischen Aussagen nicht: Auch jener Begriff von Wahrheit, der Standards von Tatsachen so streng zu unterscheiden erlaubt, ist, wie immer wir uns an ihm bloß unausdrücklich orientieren, doch seinerseits ein Standard, der kritischer Rechtfertigung bedarf."
11. See Albert (1969, p. 233): "Es dürfte kaum sinnvoll sein, alle solche Unterscheidungen in einer ad hoc postulierten Einheit von Vernunft und Entscheidung >>dialektisch aufzuheben<< und damit die verschiedenen Aspekte von Problemen und die Ebenen der Argumentation in einer Totalität untergehen zu lassen, die zwar alles gleichzeitig umfaßt, aber auch dazu nötigt, alle Probleme gleichzeitig zu lösen. Ein solches Verfahren kann nur dazu führen, daß Probleme angedeutet, aber nicht mehr analysiert, Lösungen prätendiert, aber nicht durchgeführt werden. Der dialektische Kult der totalen Vernunft ist zu anspruchsvoll, um sich mit >>partikularen<< Lösungen zu begnügen. Da es keine Lösungen gibt, die seinen Ansprüchen genügen, ist er

genötigt, sich mit Andeutungen, Hinweisen und Metaphern zufrieden zu geben."

12. See Albert (1963/1993): "Die in wissenschaftlichen Theorien beschriebenen Gesetzmäßigkeiten können als Spielräume angesehen werden, innerhalb deren das tatsächliche Geschehen abläuft. Sie schließen also bestimmte logisch mögliche Vorgänge und Ereignisse (bzw. Häufigkeitsverteilungen von Ereignissen) aus. Ihre praktische Relevanz liegt darin, daß sie die menschlichen Wirkungsmöglichkeiten in bestimmter Weise einschränken, gewissermaßen >>kanalisieren<<, also den Spielraum der Handlungsmöglichkeiten festlegen und damit eine Antwort auf die Frage ermöglichen: *Was können wir tun?* Um diese Antwort explizit zu machen, kann man theoretische Systeme in eine technologische Form bringen, in der das mögliche Geschehen auf mögliche Ansatzpunkte für menschliches Handeln bezogen wird. Dadurch werden Möglichkeiten des tatsächlichen Geschehens in menschliche Handlungsmöglichkeiten transformiert, unter Berücksichtigung der Tatsache, daß an bestimmten Stellen in den Ablauf praktisch eingegriffen wird. Auch die *Technologie* enthält also an sich keine Vorschriften, sondern Feststellungen: *informative* Aussagen."
13. See Habermas (1965, p. 155): "Für drei Kategorien von Forschungsprozessen läßt sich ein spezifischer Zusammenhang von logisch-methodologischen Regeln und erkenntnisleitenden Interessen nachweisen [...]. In den Ansatz der empirisch-analytischen Wissenschaften geht ein *technisches*, in den Ansatz der historisch-hermeneutischen Wissenschaften ein *praktisches* und in den Ansatz der kritisch orientierten Wissenschaften jenes *emanzipatorisches* Interesse ein, das schon den traditionellen Theorien uneingestanden [...] zugrunde lag." For a critical discussion, see Keuth (1989, p. 180ff.), and for a friendly appraisal, see Honneth (2017).
14. See Poincaré (1919, p. 225).
15. See Duhem (1906/1981, p. 221f.): "Une expérience de Physique est l'observation précise d'un groupe de phénomènes accompagnée de l'INTERPRÉTATION de ces phènomènes; cette interpretation substitue aux données concrètes réellement recueillies par l'observation des representations abstraites et symboliques qui leur correspondent en vertu des théories admises par l'observateur."

16. See Duhem (1906/1981, p. 230): "Un fait théorique unique peut donc se traduire par une infinité de faits pratiques disparates; un fait pratique unique correspond à une infinité de faits théoriques incompatibles; cette double constatation fait éclater aux yeux la verité que nous voulions mettre en évidence: Entre les phénomènes réellement constatés au cours d'une experience et le résultat de cette experience, formulée par le physicien, s'intercale une élaboration intellectuelle très complexe qui, à un récit de faits concrets, substitute un jugement abstrait et symbolique."
17. See Duhem (1906/1981, p. 388): "La contemplation d'un ensemble de lois d'expérience ne suffit donc pas à suggérer au physicien quelles hypothèses il doit choisir pour donner, de ces lois, une représentation théorique; il faut encore que les pensées habituelles à ceux au milieu desquels il vit, que les tendances imprimées à son propre esprit par ses études antérieures viennent le guider et restreindre la latitude trop grande que les règles de la Logique laissent à ses démarches. Combien de parties de la Physique gardent, jusqu'à ce jour, la forme purement empirique, attendant que les circonstances préparent la génie d'un physicien à concevoir les hypothèses qui les organiseront en théorie!"
18. Duhem refers to Pascal's thought about truth without referring to justice which makes up the same thought of Blaise Pascal (1670/2000, p. 72f.): "La justice et la verité sont deux points si subtiles que nos instruments sont trop émoussés pour y toucher exactement. S'ils y arrivent, ils en écachent la pointe, et appuient tout autour plus sur le faux que sur le vrai."
19. See also Duhem (1908/2003, p. 17).
20. See Duhem (1906/1981, p. 332): "La saine critique expérimentale d'une hypothèse est subordonnée à certaines conditions morales; pour apprécier exactement l'accord d'une théorie physique avec les faits, il ne suffit pas d' être bon géomètre et expérimentateur habile, il faut encore être juge impartial et loyal."
21. This interpretation seems to be supported by what Duhem writes in his La Science Allemande (1915, p. 56): "Dans le domaine de toute de science, mais, plus encore, dans le domaine d l'histoire, la recherche de la verité ne requiert pas seulement des aptitudes intellectuelles; elle réclame en outre des qualités morales, la droiture, la probité, le détachement de tout intérèt et de toute passion."

    An interesting interpretation of Duhem's reflections of the importance of *bon sens* for theory choice as an early case of virtue epistemology, see Stump (2007, 2011) and the subsequent debate: Ivanova (2010), Kidd (2011), Ivanova (2011), Ivanova and Paternote (2013) and Dietrich and Honenberger (2020).

22. See Levi (1960, p. 356): "[It is not the case] that the scientist *qua* scientist makes no value judgment but that given his commitment to canons of inference he need make no further value judgments in order to decide which hypotheses to accept and which to reject." See also Levi (1962).
23. Richard Jeffrey (1956, p. 242) pointed at the complexity of the decision to reject or accept a hypothesis that the scientists had to face and that it would be unreasonable to expect that scientists could possibly consider all the consequences of such a decision: "It is certainly meaningless to speak of *the* cost of mistaken acceptance or rejection, for by its nature a putative scientific law will be relevant in a great diversity of choice situations among which the cost of a mistake will vary greatly."
24. See Hempel (1965, p. 93): "But in a general way, it seems clear that the standards governing the inductive procedures of pure science reflect the objective of obtaining a certain goal, which might be described somewhat vaguely as the attainment of an increasingly reliable, extensive, and theoretically systematized body of information about the world. Note that if we were concerned, instead, to form a system of beliefs or a world view that is emotionally reassuring or aesthetically satisfying to us, then it would not be reasonable to insist, as science does, on a close accord between the beliefs we accept and our empirical evidence; and the standards of objective testability and confirmation by publicly ascertainable evidence would have to be replaced by acceptance standards of an entirely different kind. The standards of procedure must in each case be formed in consideration of the goals to be attained; their justification must be relative to those goals and must, in this sense, presuppose them."
25. See also his remark in *The Open Society and Its Enemies* (1945/2002, p. 511): "To be sure, it is impossible to prove the rightness of any ethical principle, or even to argue in its favour in just the manner in which we argue in favour of a scientific statement. Ethics is not science. But although there is no 'rational scientific basis' of ethics, there is an ethical basis of science, and of rationalism."
26. See Nagel (1961, p. 488f.).

27. See Nagel (1961, p. 489): "Although the recommendation that social scientists make fully explicit their value commitments is undoubtedly salutary, and can produce excellent fruit, it verges on being a counsel of perfection. For the most part we are unaware of many assumptions that enter into our analyses and actions, so that despite resolute efforts to make our preconceptions explicit some decisive ones may not even occur to us. But in any event, the difficulties generated for scientific inquiry by unconscious bias and tacit value orientations are rarely overcome by devout resolutions to eliminate bias. They are usually overcome, often only gradually, through the self-correcting mechanisms of science as a social enterprise. For modern science encourages the invention, the mutual exchange, and the free but responsible criticisms of ideas; it welcomes competition in the quest for knowledge between independent investigators, even when their intellectual orientations are different; and it progressively diminishes the effects of bias by retaining only those proposed conclusions of its inquiries that survive critical examination by an indefinitely large community of students, whatever be their value preferences or doctrinal commitments. It would be absurd to claim that this institutionalized mechanism for sifting warranted beliefs has operated or is likely to operate in social inquiry as effectively as it has in the natural sciences. But it would be no less absurd to conclude that reliable knowledge of human affairs is unattainable merely because social inquiry is frequently value-oriented."
28. See Kuhn (1962/1970, p. 199f.): "There is no neutral algorithm for theory-choice, no systematic decision procedure which, properly applied, must lead each individual in the group to the same decision. In this sense it is the community of specialists rather than its individual members that makes the effective decision. To understand why science develops as it does, one need not unravel the details of biography and personality that lead each individual to a particular choice, though that topic has vast fascination. What one must understand, however, is the manner in which a particular set of shared values interacts with the particular experiences shared by a community of specialists to ensure that most members of the group will ultimately find one set of arguments rather than another decisive."
29. See Kuhn (1977, p. 321ff.).

30. See also Lakatos's important remarks on rationality (1970, p. 174f.): "In the light of my considerations, the idea of instant rationality can be seen to be utopian. But this utopian idea is a hallmark of most brands of epistemology. Justificationists wanted scientific theories to be proved even before they were published; probabilists hoped a machine could flash up instantly the value (degree of confirmation) of a theory, given the evidence; naïve falsificationists hoped that elimination at least was the instant result of the verdict of *experiment*. I hope I have shown that *all these theories of instant rationality – and instant learning – fail.* The case studies of this section show that rationality works much slower than most people tend to think, and, even then, fallibly. Minerva's owl flies at dusk. I also hope I have shown that the *continuity* in science, the *tenacity* of some theories, the rationality of a certain amount of dogmatism, can only be explained if we construe science as a battleground of research programmes rather than of isolated theories."
31. See his nineteenth thesis in his *Against Method* (1975/2010, p. 249): "Science is neither a single tradition, nor the best tradition there is, except for people who have become accustomed to its presence, its benefits and its disadvantages. In a democracy it should be separated from the state just as churches are now separated from the state."
32. See also his *Second Dialogue on Knowledge* (1976/1991, p. 113): "Stone Age man was already the fully developed *homo sapiens*, he was faced by tremendous problems, and he solved them with great ingenuity. Science is always praised because of its achievements. So let us not forget that the inventors of myth invented fire, and the means of keeping it. They domesticated animals, bred new types of plants [...]. If science is praised because of its achievements, then myth must be praised a hundred times more fervently because its achievements were incomparably greater: the inventors of myth started culture while scientists just changed it, and not always for the better. I have already mentioned one example: myth, tragedy, the older epics dealt with emotions, fact, structures all at the same time, and they had a profound and beneficial influence on the societies in which they occurred."
33. See, e.g., Rorty (1979/2018, p. 178f.): "To say that the True and the Right are matters of social practice may seem to condemn us to a

relativism which, all by itself, is a *reductio* of a behaviorist approach to either knowledge or morals. [...] I shall simply remark that only the image of a discipline – philosophy – which will pick out a given set of scientific or moral views as more 'rational' than the alternatives by appeal to something which forms a permanent neutral matrix for all inquiry and all history, makes it possible to think that such relativism must automatically rule out coherence theories of intellectual and practical justification. One reason why professional philosophers recoil from the claim that knowledge may not have foundations, or rights and duties an ontological ground, is that the kind of behaviorism which dispenses with foundations is in a fair way toward dispensing with philosophy. For the view that there is no permanent neutral mix within which the dramas of inquiry and history are enacted has as a corollary that criticism of one's culture can only be piecemeal and partial – never 'by reference to eternal standards.' This threatens the neo-Kantian image of philosophy's relation to science and to culture. The urge to say that assertions and actions must not only cohere with other assertions and actions but 'correspond' to something apart from what people are saying and doing has some claim to be called *the* philosophical urge." However, the "value-free ideal" has continued to be defended, see, e.g., McMullin (1983/2012).

34. See Gross and Levitt (1994). An important incident was Sokal's hoax (1996): an article that the physicist submitted to a journal of postmodern critical theory, *Social Text*, entitled "Transgressing the Boundaries: Towards a Transformative Hermeneutics of Quantum Gravity" in order to test the journal's rigor. The article deliberately included nonsense but was accepted for publication. Sokal revealed in the magazine *Lingua Franca* that the article was a hoax. He later published with Bricmont the book *Impostures Intellectuelles* (1997). See also the contributions in Koertge (1998).
35. See Laudan (1984, p. 61f.): "More generally, there are plenty of cases of axiological disagreement in which there is ample scope for fully rational individuals to disagree about goals even when they fully agree about shared examples. But that is a far cry from the familiar claim [...] that virtually all cases of disagreement about cognitive values are beyond rational resolution. It is crucial for us to understand that scientists do sometimes change their minds

about their most basic cognitive ends, and sometimes they can give compelling arguments outlining the reasons for such changes. In this regard, disagreements about goals are exactly on a par with factual and methodological disputes. Sometimes they can be rationally brought to closure; other times, they cannot. But there is nothing about the nature of cognitive goals which makes them intrinsically immune to criticism and modification."

36. See also Laudan's dialogue on relativism (1990) and Laudan (2004).
37. See also her definition of knowledge in Longino (1990, p. 185f.): "What we know is what we can experience. The conclusions of inferences from experience that must use additional substantive assumptions as premises cannot be known absolutely. We give the name 'knowledge' to the complex and more or less coherent sets of hypotheses, theories, and experimental-observational data accepted by a culture at a given time because this body of ideas functions as a public fund of justification and legitimization for new hypotheses as well as for action and policy. This socially created knowledge which integrates experience and the needs and assumptions of a culture is true relative to those assumptions and, to the extent that those assumptions are context-dependent, is relative to that context. If scientific knowledge is social knowledge, to hold scientific claims to strict empirical criteria is to remain agnostic with respect to the context-independent truth or falsity of many of them." Is this really different from relativism?
38. See the devastating criticism of Keuth (1993) in his book *Erkenntnis oder Entscheidung?*
39. Putnam (1981) has made a similar point already in his *Reason, Truth and History* (p. 133f.): "What I have been saying is that the procedures by which we decide on the acceptability of a scientific theory have to do with whether or not the scientific theory as a whole exhibits certain 'virtues'. I am assuming that the procedure of building up scientific theory cannot be correctly analyzed as a procedure of verifying scientific theories *sentence by sentence*. I am assuming that verification in science is a holistic matter, that it is whole theoretical systems that meet the test of experience 'as a corporate body', and that the judgment of how well a whole system of sentences meets the test of experience is ultimately somewhat of an intuitive matter which

could not be formalized short of formalizing total human psychology. […] The fact is that, if we consider the ideal of rational acceptability which is revealed by looking at what theories scientists and ordinary people consider rational to accept, then we see that what we are trying to do in science is to construct a representation of the world which has the characteristics of being instrumentally efficacious, coherent, comprehensive, and functionally simple. But why? I would answer that the reason we want this sort of representation […] is that having this sort of representation system is *part of our idea of human cognitive flourishing*, and hence part of our idea of total human flourishing, of Eudaemonia."

40. It was Bernard Williams (1985) who introduced and analysed "thick concepts."
41. See Keuth (1989, p. 28): "Weber betont erneut, daß die Produktion wissenschaftlicher Tatsachenbehauptungen keineswegs *von allen Wertungen unabhängig* ist. Vielmehr weisen 'Kultur- und das heißt *Wert*interessen … auch der rein empirisch-wissenschaftlichen Arbeit die Pfade' [Weber 1923/1996, p. 168]. Man muß sich aber nicht mit dieser Tatsachenbehauptung begnügen, denn es läßt sich leicht zeigen, woran die Unabhängigkeit scheitert. Jede Äußerung ist ja eine Sprechhandlung. Doch zu einer solchen *Handlung* genügt das verbale Verhalten der Produktion einer Aussage nicht. Vielmehr muß der Sprecher auch *beabsichtigen*, sich so zu verhalten. Er muß also diese Verhaltensweise allen anderen, die er an ihrer Stelle zeigen könnte, und der Untätigkeit *vorziehen*, d.h. er muß sie gegenüber diesen Alternativen höher *bewerten*. Aber dadurch, daß eine solche Wertung Voraussetzung der Äußerung einer Aussage ist, wird die Aussage selbst nicht zum Werturteil. Das hätte auch die absurde Konsequenz, daß wir stets Werturteile produzieren, selbst wenn wir nur etwas behaupten oder fragen wollen."
42. Magnus (2013, p. 845) claims that the argument from inductive risk is still older, going back to William James's essay on "The Will to Believe" (1896). For further references on the current discussion on inductive risk, see footnote 4 of chapter 2.
43. In the current discussion, extreme relativistic tendencies are seldom. Even sociologists of sciences, such as Collins (2023, p. 297) who

conceptualizes science as 'craftwork with integrity', acknowledge the importance of truth: "The science we should love will be driven by the search for truth even if the purity of the motive is hard to maintain because the forces that subvert it are unnoticeable. But the science we should love will never knowingly subvert the formative motive; when it is shown that the motive is unattainable, the science we should love will still aspire to attain it. The same applies to truth itself: the truth may not be found, but the goal must be to find it. The reason to love science in spite of the unattainability of its goals is that, unlike so many institutions that run on greed and the quest for power, it is an institution in which truth, and therefore integrity, are foundational." See Lacey (1999). Elliott (2017) and Elliott (2022) are introductory textbooks on science and values. Kincaid, Dupré and Wylie (2007), Schurz and Carrier (2013) and Elliott and Steel (2017) are edited volumes on the Value-Free-Ideal. Bright (2018) discusses Du Bois democratic defence of the value-free ideal from the point of view of the contemporary discussion.

References

Acierno, Louis (1994): *The History of Cardiology*, London and New York: The Parthenon Publishing Group.

Adams, Marcus (2009): "Empirical Evidence and the Knowledge-That/Knowledge-How Distinction", *Synthese*, vol. 170, pp. 97–114.

Adorno, Theodor et al. (eds.) (1969): *Der Positivismusstreit in der deutschen Soziologie*, Neuwied, Berlin: Hermann Luchterhand Verlag.

Albert, Hans (1963/1993): "Wertfreiheit als methodisches Prinzip", in Erwin von Beckerath and Herbert Giersch (eds.): *Probleme der Normativen Ökonomik und der Wirtschaftspolitischen Beratung*, Berlin: Duncker & Humblot, 1963, pp. 32–63 and reprinted in Topitsch, Ernst (ed.): *Logik der Sozialwissenschaften*, 12th ed., Athenäum, Hain, Hanstein: Neue wissenschaftliche Bibliothek, 1993, pp. 196–225.

Albert, Hans (1966): "Theorie und Praxis: Max Weber und das Problem der Wertfreiheit und der Rationalität", in Ernst Oldemeyer (ed.): *Die Philosophie und die Wissenschaften, Simon Moser zum 65. Geburtstag*, Meisenheim, pp. 246–272 and reprinted in Albert and Topitsch (1979), pp. 200–236.

Albert, Hans (1968/1985): *Treatise on Critical Reason*, Princeton: Princeton University Press.

Albert, Hans (1969): "Der Mythos der totalen Vernunft: Dialektische Ansprüche im Lichte undialektischer Kritik", in Adorno et al. (eds.): (1969), pp. 193–234.

Albert, Hans (1975): *Transzendentale Träumereien*, Hamburg: Hoffmann und Campe.

Albert, Hans (1982): *Die Wissenschaft und die Fehlbarkeit der Vernunft*, Tübingen: J.C.B. Mohr (Paul Siebeck).

Albert, Hans (1987): *Kritik der reinen Erkenntnislehre*, Tübingen: J.C.B. Mohr (Paul Siebeck).

Albert, Hans (2010): "The Economic Tradition and the Constitution of Science", *Public Choice*, vol. 144, pp. 401–411.

Albert, Hans and Topitsch, Ernst (eds.) (1979): *Werturteilsstreit*, Darmstadt: Wissenschaftliche Buchgesellschaft.

Albert, Max (2011): "Methodology and Scientific Competition", *Episteme*, vol. 8, pp. 165–183.

Albert, Max, Schmidtchen, Dieter, and Voigt, Stefan (eds.) (2008): *Scientific Competititon*, Tübingen: J.C.B. Mohr Siebeck.

Alt, James and Crystal, K. Alec (1983): *Political Economics*, Berkeley: University of California Press.

Anderson, Elisabeth (2004): "Uses of Value Judgments in Science: A General Argument, with Lessons from a Case Study of Feminist Research on Divorce", *Hypatia*, vol. 19, pp. 1–24.

Anderson, John R. (2010): *Cognitive Psychology and Its Implications*, 7th ed., New York: Worth Publishers.

Anderson, Lanier (2003): "The Debate over the *Geisteswissenschaften* in German *Philosophy*, 1880–1910", in Thomas Baldwin (ed.): *The Cambridge History of Philosophy: 1870–1945*, Cambridge: Cambridge University Press, pp. 221–234.

Andrews, Kristin (2016): "Animal Cognition", in Edward N. Zalta (ed.): *Stanford Encyclopedia of Philosophy* (Summer 2016 Edition), https://plato.stanford .edu/archives/sum2016/entries/cognition-animal/.

Apel, Karl-Otto (1973): "Das Apriori der Kommunikationsgemeinschaft und die Grundlagen der Ethik: Zum Problem einer rationale Begründung der Ethik im Zeitalter der Wissenschaft", in *Transformation der Philosophie, Bd. 2, Das Apriori der Kommunikationsgesellschaft*, Frankfurt am Main: Suhrkamp, pp. 358–435.

Apel, Karl-Otto (1976): "Sprechakttheorie und transzendentale Sprachpragmatik zur Frage ethischer Normen", in *Sprachpragmatik und Philosophie*, Frankfurt am Main: Suhrkamp, pp. 10–173.

Arabatzis, Theodore (2006): "On the Inextricability of the Context of Discovery and the Context of Justification", in Jutta Schickore and Friedrich Steinle (eds.): *Revisiting Discovery and Justification*, Dordrecht: Springer, pp. 215–230.

Arabatzis, Theodore (2017): "What's in It for the Historian of Science? Reflections on the Value of Philosophy of Science for History of Science", *International Studies in the Philosophy of Science*, vol. 31, pp. 69–82.

Arabatzis, Theodore and Kindi, Vasso (2013): "The Problem of Conceptual Change in the Philosophy and History of Science", in Stella Vosniadou (ed.): *International Handbook of Research on Conceptual Change*, London and New York: Routledge, pp. 343–359.

Aristotle, (1957): *Metaphysica*, (ed.) Werner Jaeger, Oxford: Oxford University Press.

Aristotle, (1996): *Politics*, (ed.) Stephen Everson, *Politics and The Constitution of Athens*, Cambridge: Cambridge University Press.

Austin, John L. (1962): *How to Do Things with Words*, Oxford: Clarendon Press.

Ayer, Alfred Jules (1936/1952): *Language, Truth and Logic*, 2nd ed., New York: Dover Edition.

Axelrod, Robert (1986): "An Evolutionary Approach to Norms", *American Political Science Review*, vol. 80, pp. 1095–1111.

Bacon, Francis (1620/2000): *The New Organon*, (eds.) Lisa Jardine, and Michael Silverthorne, Cambridge: Cambridge University Press.

Bandura, Albert (1986): *Social Foundations of Thought and Action: A Social Cognitive Theory*, Englewood Cliffs, NJ: Prentice-Hall.

Bargh, John and Chartrand, Tanya (1999): "The Unbearable Automaticity of Being", *American Psychologist*, vol. 54, pp. 461–479.

Bechtel, William and Abrahamsen, Adele (2005): "Explanation: A Mechanistic Alternative", *Studies in History and Philosophy of Biology and Biomedical Sciences*, vol. 36, pp. 421–441.

Bernholz, Peter, Streit, Manfred E., and Vaubel, Roland (1998): *Political Competition, Innovation and Growth*, Berlin and New York: Springer.

Bicchieri, Cristina (2005): *The Grammar of Society*, Cambridge: Cambridge University Press.

Bicchieri, Cristina (2016): *Norms in the Wild*, Oxford: Oxford University Press.

Biddle, Justin (2013): "State of the Field: Transient Underdetermination and Values in Science", *Studies in History and Philosophy of Science*, vol. 44, pp. 124–133.

Bird, Alexander (2007): "What Is Scientific Progress?", *Noûs*, vol. 41, pp. 64–89.

Bird, Alexander (2016): "Scientific Progress", in Paul Humphreys (ed.): *The Oxford Handbook of Philosophy of Science*, Oxford: Oxford University Press, pp. 544–563.

Bird, Alexander (2019): "The Aim of Belief and the Aim of Science", *Theoria: An International Journal for Theory, History and Foundations of Science*, vol. 34, pp. 171–193.

Bird, Alexander (2023): "The Epistemic Approach: Scientific Progress as Accumulation of Knowledge", in Yafeng Shan, (ed.): *New Philosophical Perspectives on Scientific Progress*, New York and London: Routledge, pp. 13–26.

Blackburn, Simon and Simmons, Keith (eds.) (1999): *Truth*, Oxford: Oxford University Press.

Blackburn, Simon (1971/1993): "Moral Realism" in J. Casey (ed.): *Morality and Moral Reasoning*, London: Methuen, 1971, reprinted in *Essays on Quasi-Realism*, Oxford: Oxford University Press, 1993, pp. 111–123.

Blackburn, Simon (1988): *Ruling Passions: A Theory of Practical Reason*, Oxford: Clarendon Press.

Blackburn, Simon (1993): *Essays on Quasi-Realism*, Oxford: Clarendon Press.

Blaug, Mark (1997): *Economic Theory in Retrospect*, 5th ed., Cambridge: Cambridge University Press.

Bloor, David (1976/1991): *Knowledge and Social Imagery*, 2nd ed., Chicago: The University of Chicago Press.

Böckenförde, Ernst-Wolfgang (1976): *Staat, Gesellschaft, Freiheit: Studien zur Staatstheorie und zum Verfassungsrecht*, Frankfurt am Main: Suhrkamp.

Bokulich, Alisa (2018): "Representing and Explaining: The Eikonic Conception of Scientific Explanation", *Philosophy of Science*, vol. 85, pp. 793–805.

Bouterse, Jeroen and Bart Karstens (2015): "A Diversity of Divisions: Tracing the History of the Demarcation between the Sciences and the Humanities", *Isis*, vol. 106, pp. 341–352.

Boyd, Richard (1988): "How to Be a Moral Realist", in Geoffrey Sayre-McCord (ed.): *Essays on Moral Realism*, Ithaca: Cornell University Press, pp. 187–228.

Boyd, Robert and Richerson, Peter J. (1985): *Culture and the Evolutionary Process*, Chicago: The University of Chicago Press.

Boyd, Robert and Richerson, Peter J. (2009): "Culture and the Evolution of Human Cooperation", *Philosophical Transactions of the Royal Society B*, vol. 364, pp. 3281–3288.

Boyer-Kassem, Thomas, Mayo-Wilson, Conor, and Weisberg, Michael (eds.) (2018): *Scientific Collaboration and Collective Knowledge: New Essays*, Oxford: Oxford University Press.

Brennan, Geoffrey and Buchanan, James (1985): *The Reason of Rules: Constitutional Political Economy*, Cambridge: Cambridge University Press.

Bresson, Alain (2016): *The Making of the Ancient Economy: Institutions, Markets, and Growth in the City States*. Princeton: Princeton University Press.

Brigandt, Ingo (2015): "Social Values Influence the Adequacy Conditions of Scientific Theories: Beyond Inductive Risk", *Canadian Journal of Philosophy*, vol. 45, pp. 326–356.

Bright, Liam Kofi (2017): "On Fraud", *Philosophical Studies*, vol. 174, pp. 291–310.

Bright, Liam Kofi (2018): "Du Bois' Democratic Defence of the Value Free Ideal", *Synthese*, vol. 195, pp. 2227–2245.

Brink, David (1989): *Moral Realism and the Foundations of Ethics*, Cambridge: Cambridge University Press.

Brown, Matthew J. (2013): "Values in Science beyond Underdetermination and Inductive Risk", *Philosophy of Science*, vol. 80, pp. 829–839.

Buchanan, James (1975): *The Limits of Liberty. Between Anarchy and Leviathan*, Chicago: The University of Chicago Press.

Buchanan, James (1989): "The Relatively Absolute Absolutes", in James Buchanan (ed.): *Essays in Political Economy*, Honolulu: University of Hawaii Press, pp. 96–109.

Buchanan, James (2001): "Law and the Invisible Hand", in *Moral Science and Moral Order, the Collected Works of James M. Buchanan Vol. 17*, Indianapolis: The Liberty Press, pp. 96–109.

Buchanan, James and Tullock, Gordon (1962): *The Calculus of Consent: Logical Foundations of Constitutional Democracy*, Ann Arbor: University of Michigan Press.

Bueter, Anke (2015): "The Irreducibility of Value-Freedom to Theory Assessment", *Studies in History and Philosophy of Science*, vol. 49, pp. 18–26.

Burke, Edmund (1791/1962): *An Appeal from the New to the Old Whigs*, ed. J. M. Robson, Indianapolis, IN: Bobbs-Merrill.

Campbell, Donald (1965): "Variation and Selective Retention in Socio-Cultural Evolution", in Herbert R. Barringer, George I. Blanksten, and Raymond W. Mack (eds.): *Social Change in Developing Areas*, Cambridge, MA: Schenkman, pp. 19–49.

Capella, Martianus (1977): *The Marriage of Philology and Mercury*, (eds.) Stahl, William Harris and Burge, E. L., New York: Columbia University Press.

Carrier, Martin (2013): "Values and Objectivity in Science: Value-Ladenness, Pluralism and the Epistemic Attitude", *Science & Education*, vol. 22, pp. 2547–2568.

Carrier, Martin (2017): "Facing the Credibility Crisis of Science: On the Ambivalent Role of Pluralism in Establishing Relevance and Reliability", *Perspectives on Science*, vol. 25, pp. 439–464.

Carrier, Martin (2018): "Identifying Agnotological Ploys: How to Stay Clear of Unjustified Dissens", in A. Christian et al. (eds.): *Philosophy of Science – Between the Natural Sciences, the Social Sciences, and the Humanities*, New York: Springer, pp. 155–169.

Carrier, Martin (2022): "What Does Good Science-Based Advice to Politics Look Like?", *Journal for General Philosophy of Science*, vol. 53, pp. 5–21.

Chakravartty, Anjan (2017): *Scientific Ontology: Integrating Naturalized Metaphysics and Voluntarist Epistemology*, Oxford. Oxford University Press.

Chang, Hasok (2012): *Is Water H_2O? Evidence, Realism and Pluralism*, Berlin and New York: Springer.

ChoGlueck, Christopher (2018): "The Error Is in the Gap: Synthesizing Accounts for Societal Values in Science", *Philosophy of Science*, vol. 85, pp. 704–725.

Clauberg, Johannes (1654): *Logica, Vetus & Nova*, Amsterdam: Ex Officina Elzeviriana.

Coase, Ronald (1937): "The Nature of the Firm", *Economica*, vol. 4, pp. 386–405.

Coase, Ronald (1960): "The Problem of Social Cost", *Journal of Law and Economics*, vol. 3, pp. 1–44.

Cohen, Floris (2015): *The Rise of Modern Science Explained: A Comparative History*, Cambridge: Cambridge University Press.

Coleman, James (1990a): *Foundations of Social Theory*, Cambridge: Harvard University Press.

Coleman, James (1990b): "The Emergence of Norms", in Michael Hechter, Karl-Dieter Opp, and Reinhard Wippler (eds.): *Social Institutions*, Berlin and New York: Walter de Gruyter, pp. 35–59.

Collins, Harry (2023): "Science as Craftwork with Integrity", in David Ludwig, Inkari Koskinen, Zinhle Mncube, Luana Poliseli, and Luis Reyes-Galindo (eds.): *Global Epistemologies and Philosophies of Science*, London and New York: Routledge, pp. 296–307.

Condorcet, Marie Jean Antoine Nicolas de Caritat (1785): *Essai sur l' Application de l' Analyse à la Probabilité des Décisions Rendues à la Pluralité des Voix*, Paris: L' Imprimerie Royale, Facsimile edition: New York, Chelsea, 1972.

Contessa, Gabriele (2021): "On the Mitigation of Inductive Risk", *European Journal for Philosophy of Science*, vol. 11, p. 64, https://doi.org/10.1007/s13194-021-00381-6.

Congleton, Roger (1989): "Efficient Status Seeking: Externalities, and the Evolution of Status Games", *Journal of Economic Behavior and Organization*, vol. 11, pp. 175–190.

Cooper, John (1997): *Plato, Complete Works*, Indianapolis: Hackett Publishing.

Copp, David (2015): "Explaining Normativity", *Proceedings and Addresses of American Philosophical Association*, vol. 89, 2015, pp. 48–73.

Craver, Carl (2007): *Explaining the Brain*, Oxford: Oxford University Press.

Craver, Carl (2014): "The Ontic Account of Scientific Explanation", in Marie Kaiser et al. (eds.): *Explanation in the Special Sciences: The Case of Biology and History*, Dordrecht: Springer, pp. 27–52.

D' Andrade, Roy (1995): *The Development of Cognitive Anthropology*, Cambridge: Cambridge University Press.

Damasio, Antonio (1994): *Descartes' Error: Emotion, Reason and the Human Brain*, New York: Avon Books.

Damasio, Antonio (1999): *The Feeling of What Happens: Body and Emotion in the Making of Consciousness*, New York: Harcourt Brace & Company.

Daston, Lorraine (1992): "Objectivity and the Escape from Perspective", *Social Studies of Science*, vol. 22, pp. 597–618.

Daston, Lorraine (2014): "Objectivity and Impartiality: Epistemic Virtues in the Humanities", in *The Making of the Humanities, Vol. III, The Modern Humanities*, Amsterdam: Amsterdam University Press, pp. 27–41.

Daston, Lorraine and Galison, Peter (2007): *Objectivity*, New York: Zone Books.

Davidson, Donald (1984): *Inquiries into Truth and Interpretation*, Oxford: Clarendon Press.

Davies, Martin (2015): "Knowledge (Explicit, Implicit and Tacit): Philosophical Aspects", in James D. Wright (ed.): *International Encyclopedia of the Social and Behavioral Sciences*, 2nd ed., vol. 13, London: Elsevier, pp. 74–90.

Deaton, Angus and Cartwright, Nancy (2018): "Understanding and Misunderstanding Randomized Controlled Experiments", *Social Science & Medicine*, vol. 210, pp. 2–21.

De Maistre, Joseph (1794–1795/1965): "Study on Sovereignty", in J. Lively (ed.): *Works*, New York: Macmillan, pp. 93–130.

De Melo-Martín, Inmaculada and Intemann, Kristen (2016): "The Risk of Using Inductive Risk to Challenge the Value-Free Ideal", *Philosophy of Science*, vol. 83, pp. 500–520.

Dellsén, Finnur (2021): "Understanding Scientific Progress: The Noetic Account", *Synthese*, vol. 199, pp. 11249–11278.

Dellsén, Finnur (2023): "The Noetic Approach: Scientific Progress as Enabling Understanding", in Yafeng Shan (ed.): *New Philosophical Perspectives on Scientific Progress*, New York and London: Routledge, pp. 62–81.

Denzau, Arthur and North, Douglass C. (1994): "Shared Mental Models: Ideologies and Institutions", *Kyklos*, vol. 47, pp. 3–31.

Devitt, Michael (2011): "Methodology and the Nature of Knowing How", *Journal of Philosophy*, vol. 108, pp. 205–218.

Dhami, Sanjit (2017): *The Foundations of Behavioral Economic Analysis*, Oxford: Oxford University Press.

Dietrich, Michael and Philip Honnenberger (2020): "Duhem's Problem Revisited: Logical vs. Epistemic Formulations and Solutions", *Synthese*, vol. 197, pp. 337–354.

DiMaggio, Paul (1997): "Culture and Cognition", *Annual Review of Sociology*, vol. 23, pp. 263–287.

DiMaggio, Paul and Powell, Walter (eds.) (1991): *The New Institutionalism in Organizational Analysis*. Chicago: Chicago University Press.

Dilthey, Wilhelm (1883/1990): *Gesammelte Schriften, I. Band: Einleitung in die Geisteswissenschaften*, 9. Auflage, Stuttgart: B.G. Teubner Verlagsgesellschaft und Göttingen: Vandenhoeck & Ruprecht.

Dilthey, Wilhelm (1924/1990): *Gesammelte Schriften V. Band: Die geistige Welt. Einleitung in die Philosophie des Lebens. Erste Hälfte: Abhandlungen zur Grundlegung der Geisteswissenschaften*, 8. Auflage, Stuttgart: B.G. Teubner Verlagsgesellschaft und Göttingen: Vandenhoeck & Ruprecht.

Dilthey, Wilhelm (1927/1992): *Gesammelte Schriften VII. Band: Der Aufbau der geschichtlichen Welt in den Geisteswissenschaften*, 8. unv. Auflage, Stuttgart: B.G. Teubner Verlagsgesellschaft und Göttingen: Vandenhoeck & Ruprecht.

Dixit, Avinash, Skeath, Susan, and Reiley, David (2015): *Games of Strategy*. 4th ed., New York: W.W. Norton.

Donald, Merlin (1991): *The Origins of the Modern Mind: Three Stages in the Evolution of Culture and Cognition*, Cambridge, MA: Harvard University Press.

Dorato, Mauro (2004): "Epistemic and Nonepistemic Values in Science", in Peter Machamer and Gereon Wolters (eds.): *Science, Values and Objectivity*, Pittsburgh: University of Pittsburgh Press, pp. 52–77.

Dorr, Cian, Nebel, Jacob, and Zuehl, Jake (2022): "The Case for Comparability", forthcoming in *Noûs*.

Douglas, Heather (2000): "Inductive Risk and Values in Science", *Philosophy of Science*, vol. 67, pp. 559–579.

Douglas, Heather (2004): "The Irreducible Complexity of Objectivity", *Synthese*, vol. 138, pp. 453–473.

Douglas, Heather (2009): *Science, Policy, and the Value-Free Ideal*, Pittsburgh, PA: University of Pittsburgh Press.

Dressel, Markus (2022): "Inductive Risk: Does It Really Refute Value-Freedom?", *Theoria. An International Journal for Theory, History and Foundations of Science*, vol. 37, pp. 181–207.

Duhem, Pierre (1906/1981): *La Théorie Physique. Son Objet – Sa Structure*, Paris: Vrin.

Duhem, Pierre (1908/2003): *Sauver les Apparences. ΣΩΖΕΙΝ ΤΑ ΦΑΙΝΟΜΕΝΑ*, Paris: Vrin.

Duhem, Pierre (1915): *La Science Allemande*, Paris: Librairie Scientifique A. Hermann & Fils.

Dupré, John (2007): "Fact and Value", in Harold Kincaid, John Dupré, and Alison Wylie (eds.): *Value-Free Science? Ideals and Illusions*, Oxford: Oxford University Press, pp. 27–41.

Eddington, Arthur (1920): *Space, Time and Gravitation*, Cambridge: Cambridge University Press.

Elbert, George (2004): "Evidence, Logic and Moral Authority: Experience and the Erosion of Certainties in Illiterate and Literate Societies", in Martin Carrier, et al. (eds.): *Knowledge and the World: Challenges beyond the Science Wars*, Berlin: Springer, pp. 211–236.

Elgin, Catherine (2017): *True Enough*, Cambridge, MA: The MIT Press.

Elkins, Zachary, Ginsburg, Tom, and Melton, James (2009): *The Endurance of Nations Constitutions*, Cambridge: Cambridge University Press.

Ellickson, Robert (1991): *Order without Law*, Cambridge, MA: Harvard University Press.

Elliott, Kevin C. (2017): *The Tapestry of Values: An Introduction to Values in Science*, Oxford: Oxford University Press.

Elliott, Kevin C. (2022): *Values in Science*, Cambridge: Cambridge University Press.

Elliott, Kevin C. and Daniel, J. McKaughan (2009): "How Values in Scientific Discovery and Pursuit Alter Theory Appraisal", *Philosophy of Science*, vol. 76, pp. 598–611.

Elliott, Kevin C. and Daniel, J. McKaughan (2014): "Nonepistemic Values and the Multiple Goals of Science", *Philosophy of Science*, vol. 81, pp. 1–21.

Elliott, Kevin C. and Steel, Daniel (eds.) (2017): *Current Controversies in Values and Science*, New York: Routledge.

Elster, Jon (1989a): *The Cement of Society*, Cambridge: Cambridge University Press.

Elster, Jon (1989b): *Nuts and Bolts for the Social Sciences*, Cambridge: Cambridge University Press.

Elster, Jon (2012): "The Optimal Design of a Constituent Assembly", in Hélène Landemore and Jon Elster (eds.): *Collective Wisdom: Principles and Mechanisms*, Cambridge: Cambridge University Press, pp. 148–172.

Elster, Jon (2015): *Explaining Social Behavior: More Nuts and Bolts for the Social Sciences*, 2nd ed., Cambridge: Cambridge University Press.

Elster, Jon et al. (eds.) (2018): *Constituent Assemblies*, Cambridge: Cambridge University Press.

Ensminger, Jean (1992): *Making a Market: The Institutional Transformation of an African Society*. Cambridge: Cambridge University Press.

Ensminger, Jean and Henrich, Joseph (eds.) (2014): *Experimenting with Social Norms: Fairness and Punishment in Cross-Cultural Perspective*, New York: Russell Sage Publications.

Estlund, David (2008): *Democratic Authority: A Philosophical Framework*, Princeton: Princeton University Press.

Fara, Patricia (2009): *Science. A Four Thousand Year History*, Oxford: Oxford University Press.

Ferguson, Adam (1767/1966): *An Essay on the History of Civil Society*, Edinburgh: Edinburgh University Press.

Fernández Pinto, Manuela (2020): "Commercial Interests and the Erosion of Trust in Science", *Philosophy of Science*, vol. 87, pp. 1003–1013.

Feyerabend, Paul (1975/2010): *Against Method*, 4th ed., London and New York: Verso.

Feyerabend, Paul (1976/1991): "Second Dialogue", in Paul Feyerabend (eds.): *Three Dialogues on Knowledge*, Oxford: Basil Blackwell, pp. 47–123.

Feyerabend, Paul (1978): *Science in a Free Society*. London: New Left Books.

Files, Craig (1996): "Goodman's Rejection of Resemblance", *British Journal of Aesthetics*, vol. 36(4), pp. 398–412.

Fine, Arthur (1998): "The Viewpoint of No-One in Particular", *Proceedings and Addresses of the American Philosophical Association*, vol. 72, pp. 9–20.

Føllesdal, Dagfinn (1979): "Hermeneutics and the Hypothetico-Deductive Method", *Dialectica*, vol. 33, pp. 319–336.

Franco, Paul (2017): "Assertion, Nonepistemic Values, and Scientific Practice", *Philosophy of Science*, vol. 84, pp. 160–180.

Freeth, Tony, Alexander, Jones, Steele, John M., and Yanis, Bitsakis (2008): "Calendars with Olympiad and Eclipse Prediction on the Antikythera Mechanism", *Nature*, vol. 454, pp. 614–617.

Freeth, Tony and Alexander, Jones (2012): "The Cosmos in the Antikythera Mechanism", *Institute for the Study of the Ancient World (ISAW) Papers 4*, available at http://dlib.nyu.edu/awdl/isaw/isaw-papers/4/.

Freeth, Tony, Higgon, David, Dacanalis, Aris, MacDonald, Lindsay, Georgakopoulou, Myrto, and Wojcik, Adam (2021): "A Model of the Cosmos in the Ancient Greek Antikythera Mechanism", *Nature Scientific Reports*, vol. 11, p. 5821.

Friedman, Michael (1974): "Explanation and Scientific Understanding", *Journal of Philosophy*, vol. 71, pp. 5–19.

Friedman, Michael (2001): *The Dynamics of Reason*, Stanford: CSLI Publications.

Friedman, Michael (2002): "Kant, Kuhn, and the Rationality of Science", *Philosophy of Science*, vol. 69, pp. 171–190.

Friedman, Michael (2011): "Extending the Dynamics of Reason", *Erkenntnis*, vol. 75, pp. 431–444.

Frigg, Roman and Hunter, Matthew (eds.) (2010): *Beyond Mimesis and Convention. Representation in Art and Science*, Heidelberg, London and New York: Springer.

Galston, William A. (2011): "Pluralist Constitutionalism", in Ellen Frankel Paul, Miller Fred D., Jr., and Paul Jeffrey (eds.): *What Should Constitutions Do?*, Cambridge: Cambridge University Press, pp. 228–241.

Gascoigne, John (2019): *Science and the State*, Cambridge: Cambridge University Press.

Geddes, Barbara, Wright, Joseph, and Frantz, Erica (2014): "Autocratic Breakdown and Regime Transitions: A New Data Set", *Perspectives on Politics*, vol. 12, pp. 313–331.

Gemtos, Petros (2016): Methodologia ton koinonikon epistimon (Methodology of the Social Sciences), 5th enlarged edition, Athens: Papazisis.

Gentner, Derdre, Holyoak, Keith, and Kokinov, Boicho (eds.) (2001): *The Analogical Mind: Perspectives from Cognitive Science*, Cambridge, MA: The MIT Press.

Gibbard Allan (1990): *Wise Choices, Apt Feelings*, Cambridge, MA: Harvard University Press.

Gibbard Allan (2003): *Thinking How to Live*, Cambridge, MA: Harvard University Press.

Gibbard Allan (2014): *Meaning and Normativity*, Oxford: Oxford University Press.

Giere, Ronald (2006): *Scientific Perspectivism*, Chicago: The University of Chicago Press.

Gigerenzer, Gerd (2007): *Gut Feelings: The Intelligence of the Unconscious*, London: Penguin.

Gigerenzer, Gerd (2008): *Rationality for Mortals: How People Cope with Uncertainty*, Oxford: Oxford University Press.

Gigerenzer, Gerd (2021): "Axiomatic Rationality and Ecological Rationality", *Synthese*, vol. 198, pp. 3547–3564.

Gigerenzer, Gerd and Selten, Reinhard (eds.) (2001): *Bounded Rationality: The Adaptive Toolbox*, Cambridge, MA: The MIT Press.

Gigerenzer, Gerd, Hertwig, Ralph, and Thorsten, Pachur (eds.) (2011): *Heuristics: The Foundations of Adaptive Behavior*, Oxford: Oxford University Press.

Gigerenzer, Gerd, Todd, Paul, and the ABC Group (1999): *Simple Heuristics That Make Us Smart*, Oxford: Oxford University Press.

Gintis, Herbert (2009): *Game Theory Evolving*, 2nd ed., Princeton: Princeton University Press.

Glanzberg, Michael (ed.) (2018): *The Oxford Handbook of Truth*, Oxford: Oxford University Press.

Glymour, Clark (1980): *Theory and Evidence*, Princeton: Princeton University Press.

Gomperz, Heinrich (1939): *Interpretation: Logical Analysis of a Method of Historical Research*, Chicago: The University of Chicago Press.

Goodin, Robert and Spiekermann, Kai (2018): *An Epistemic Theory of Democracy*, Oxford: Oxford University Press.

Grandy, Richard (1973): "Reference, Meaning, and Belief", *Journal of Philosophy*, vol. 70(14), pp. 439–452.

Greenberg, Gabriel (2013): "Beyond Resemblance", *Philosophical Review*, vol. 122, pp. 215–287.

Gross, Paul and Levitt, Norman (1994): *Higher Superstition: The Academic Left and Its Quarrels with Science*, Baltimore and London: The John Hopkins University Press.

Grundmann, Thomas (2017): *Analytische Einführung in die Erkenntnistheorie*, 2nd ed., Berlin: De Gruyter.

Guala, Francesco (2009): *The Methodology of Experimental Economics*, Cambridge: Cambridge University Press.

Guala, Francesco (2016): *Understanding Institutions*, Princeton: Princeton University Press.

Habermas, Jürgen (1965/1969): "Erkenntnis und Interesse", in Jürgen Habermas (ed.): *Technik und Wissenschaft als "Ideologie"*, 3rd ed., Frankfurt am Main: Suhrkamp, pp. 146–168.

Habermas, Jürgen (1969a): "Analytische Wissenschaftstheorie und Dialektik. Ein Nachtrag zur Kontroverse zwischen Popper und Adorno", in Adorno et al. (eds.): (1969), pp. 155–191.

Habermas, Jürgen (1969b): "Gegen einen positivistisch halbierten Rationalismus", in Adorno et al. (eds.): (1969), pp. 235–266.

Hacking, Ian (1983): *Representing and Intervening*, Cambridge: Cambridge University Press.

Hadfield, Gillian and Weingast, Barry (2014): "Constitutions as Coordinating Devices", in Sebastian Galiani and Itai Sened (eds.): *Institutions, Property Rights, and Economic Growth: The Legacy of Douglass North*, Cambridge: Cambridge University Press, pp. 121–150.

Hahn, Frank (1982): "Reflections on the Invisible Hand", *Lloyds Bank Review*, vol. 144, pp. 1–21.

Hall, Peter A. (1986): *Governing the Economy: The Politics of State Intervention in Britain and France*, Oxford: Polity.

Hall, Peter and Taylor, Rosemary (1998): "Political Science and the Three New Institutionalisms", in Karol Soltan, Eric Uslaner, and Virginia Haufler (eds.): *Institutions and Social Order*, Ann Arbor: University of Michigan Press, pp. 15–43.

Hardin, Russell (1989): "Why a Constitution?", in B. Grofman and D. Wittman (eds.): *The Federalist Papers and the New Institutionalism*, New York: Agathon Press, pp. 100–120.

Hare, Richard (1952): *The Language of Morals*, Oxford: Clarendon Press.

Harvey, William (1628/1989): *Exercitatio Anatomica de Motu Cordis et Sanguinis in Animalibus*, English translation by Robert Willis in *The Works of William Harvey*, Philadelphia: University of Pennsylvania Press.

Hatzimoysis, Anthony (2002): "Analytical Descriptivism Revisited", *Ratio*, vol. 15, pp. 10–22.

Hatzis, Aristides (2015): "Moral Externalities: An Economic Approach to the Legal Enforcement of Morality", in Aristides and Nicholas Mercuro (eds.): *Law and Economics: Philosophical Issue and Fundamental Questions*, London and New York: Routledge, pp. 226–244.

Hayek, Friedrich A. von (1960): *The Constitution of Liberty*, London and New York: Routledge.

Hayek, Friedrich A. von (1973/1982): *Rules and Order*, Vol. I of *Law, Legislation and Liberty*, London and New York: Routledge.

Hayek, Friedrich A. von (1976/1982): *The Mirage of Social Justice*, Vol. II of *Law, Legislation and Liberty*, London and New York: Routledge.

Hayek, Friedrich A. von (1979/1982): *The Political Order of a Free People*, Vol. III of *Law, Legislation and Liberty*, London and New York: Routledge.

Heesen, Remco (2017): "Communism and the Incentive to Share in Science", *Philosophy of Science*, vol. 84, pp. 698–716.

Heiner, Ronald (1993): "The Origin of Predictable Behavior", *American Economic Review*, vol. 73, pp. 560–595.

Hempel, Carl (1942/1965): "The Function of General Laws in History", *Journal of Philosophy*, vol. 39, pp. 35–48 and reprinted in: Carl Hempel: *Aspects of Scientific Explanation*, New York: Free Press, pp. 231–243.

Hempel, Carl (1965): "Science and Human Values", in Carl Hempel (ed.): *Aspects of Scientific Explanation*, New York: Free Press, pp. 81–96.

Henschen, Tobias (2021): "How Strong Is the Argument from Inductive Risk?", *European Journal for Philosophy of Science*, vol. 11, p. 92, https://doi.org/10.1007/s13194-021-00409-x.

Hicks, Daniel J. (2014): "A New Direction for Science and Values", *Synthese*, vol. 191, pp. 3271–3295.

Hobbes, Thomas (1651/1991): *Leviathan*, Cambridge: Cambridge University Press.

Holman, Bennett and Wilholt, Torsten (2022): "The New Demarcation Problem", *Studies in History and Philosophy of Science*, vol. 91, pp. 211–220.

Holland, John, Holyoak, Kenneth, Nisbett, Richard, and Thagard, Paul (1986): *Induction: Processes of Inference, Learning and Discovery*, Cambridge, MA: The MIT Press.

Holt, Charles A. (2019): *Markets, Games, and Strategic Behavior: An Introduction to Experimental Economics*, 2nd ed., Princeton: Princeton University Press.

Holyoak, Kenneth and Thagard, Paul (1995): *Mental Leaps: Analogy in Creative Thought*, Cambridge, MA: The MIT Press.

Honneth, Axel (2017): "Is There an Emancipatory Interest? An Attempt to Answer Critical Theory's Most Fundamental Question", *European Journal of Philosophy*, vol. 25, pp. 908–920.

Howson, Colin and Urbach, Peter (2006): *Scientific Reasoning: The Bayesian Approach*, 3rd ed., Chicago: Open Court Publishing House.

Hoyningen-Huene, Paul (2023): "Objectivity, Value-Free Science, and Inductive Risk", *European Journal for Philosophy of Science*, vol. 13, https://doi.org/10.1007/s13194-023-00518-9.

Hudson, Robert (2016): "Why We Should Not Reject the Value-Free Ideal of Science", *Perspectives on Science*, vol. 24, pp. 167–191.

Hull, David L. (1988): *Science as a Process*, Chicago: The University of Chicago Press.

Hull, David (1997): "What's Wrong with Invisible-Hand Explanations?" *Philosophy of Science*, vol. 64, pp. S117–S126.

Hume, David (1748/1968): *Treatise of Human Nature*, Oxford: Oxford University Press.

Ivanova, Milena (2010): "Pierre Duhem's Good Sense as a Guide to Theory Choice", *Studies in History and Philosophy of Science*, vol. 41, pp. 58–64.

Ivanova, Milena (2011): "Good Sense in Context: A Response to Kidd", *Studies in History and Philosophy of Science*, vol. 42, pp. 610–612.

Ivanova, Milena and Paternotte, Cedric (2013): "Theory Choice, Good Sense and Social Consensus", *Erkenntnis*, vol. 78, pp. 1109–1132.

Jackson, Frank (1998): *From Metaphysics to Morals: A Defence of Conceptual Analysis*, Oxford: Oxford University Press.

Jackson, Frank and Pettit, Philip (1995): "Moral Functionalism and Moral Motivation", *Philosophical Quarterly*, vol. 45, pp. 20–40.

Jaeger, Werner (1936–1947): *Paideia*, 3 volumes, Berlin: De Gruyter.

James, William (1896): "The Will to Believe", *New World*, vol. 5, pp. 327–347.

Jardine, Nick and Frasca-Spada, Marina (1997): "Splendours and Miseries of the Science Wars", *Studies in History and Philosophy of Science*, vol. 28, pp. 219–235.

Jarvie, Ian (2001): *The Republic of Science: The Emergence of Popper's Social View of Science 1935–1945*, Atlanta: Rodopi.

Jefferson, Thomas (1789/1958): "Letter to James Madison, 6 September 1789", in Barbara B. Oberg (ed.): *The Papers of Thomas Jefferson*, vol. 15, Princeton: Princeton University Press.

Jeffrey, Richard (1956): "Valuation and Acceptance of Scientific Hypotheses", *Philosophy of Science*, vol. 22, pp. 237–246.

John, Stephen (2019): "Science, Truth and Dictatorship: Wishful Thinking or Wishful Speaking?", *Studies in History and Philosophy of Science*, vol. 78, pp. 64–72.

John, Stephen (2021): *Objectivity in Science*, Cambridge: Cambridge University Press.

Johnson-Laird, Philip (1983): *Mental Models*, Cambridge, MA: Harvard University Press.

Johnson-Laird, Philip (2006): *How We Reason*, Oxford: Oxford University Press, 2006.

Jones, Eric L. (2003): *The European Miracle*, 3rd ed., Cambridge: Cambridge University Press.

Kalligas, Paul (2016): "Platonic Astronomy and the Development of Ancient *Sphairopoiia*", *Rhizomata*, vol. 4(2), pp. 176–200.

Keuth, Herbert (1989): *Wissenschaft und Werturteil*, Tübingen: Mohr Siebeck.

Keuth, Herbert (1993): *Erkenntnis oder Entscheidung?*, Tübingen: Mohr Siebeck.

Khalifa, Kareem (2017): *Understanding, Explanation, and Scientific Knowledge*, Cambridge: Cambridge University Press.

Kidd, James (2011): "Pierre Duhem's Epistemic Aims and the Intellectual Virtue of Humility: A Reply to Ivanova", *Studies in the History and Philosophy of Science*, vol. 42, pp. 185–189.

Kincaid, Harold, Dupré, John, and Wylie, Alison (eds.) (2007): *Value-Free Science? Ideals and Illusions*, Oxford: Oxford University Press.

Kitcher, Philip (1981): "Explanatory Unification", *Philosophy of Science*, vol. 48, pp. 251–281.

Kitcher, Philip (1989): "Explanatory Unification and the Causal Structure of the World", in Philip Kitcher and Wesley Salmon (eds.): *Scientific Explanation*, volume 13 of Minnesota Studies in the Philosophy of Science, pp. 410–505.

Kitcher, Philip (1990): "The Cognitive Division of Labor", *Journal of Philosophy*, vol. 87, pp. 5–22.

Kitcher, Philip (1993): *The Advancement of Science*, Oxford: Oxford University Press.

Kitcher, Philip (2001): *Science, Truth, and Democracy*, Oxford: Oxford University Press.

Kitcher, Philip (2011a): *Science in a Democratic Society*, New York: Prometheus Books.

Kitcher, Philip (2011b): *The Ethical Project*, Cambridge, MA: Harvard University Press.

Kitcher, Philip (2021): *Moral Progress*, Oxford: Oxford University Press.

Kitcher, Philip (2023): "Scientific Progress and the Search for Truth" in Wenceslao Gonzalez (ed.): *Current Trends in the Philosophy of Science (Chapter 9)* Synthese Library, forthcoming.

Knight, Jack (1992): *Institutions and Social Conflict*, Cambridge: Cambridge University Press.

Koethe, John (2002): "Stanley and Williamson on Knowing How", *Journal of Philosophy*, vol. 99, pp. 325–328.

Koertge, Noretta (ed.) (1998): *A House Built on Sand: Exposing Postmodernist Myths About Science*, Oxford: Oxford University Press.

Koertge, Noretta (2000): "Science, Values, and the Value of Science", *Philosophy of Science*, vol. 67, pp. S45–S57.

Korsgaard, Christine (1996): *The Sources of Normativity*, Cambridge: Cambridge University Press.

Korsgaard, Christine (2003): "Realism and Constructivism in 20th Century Ethics", *Journal of Philosophical Research*, vol. 28(Supplement), pp. 99–122.

Koskinen, Inkeri (2020): "Defending a Risk Account of Scientific Objectivity", *British Journal for Philosophy of Science*, vol. 71, pp. 1187–1207.

Koskinen, Inkeri (2022): "How Institutional Solutions Meant to Increase Diversity in Science Fail", *Synthese*, forthcoming.

Koskinen, Inkeri (2023): "Participation and Objectivity", *Philosophy of Science*, forthcoming, https://doi.org/10.1017/psa.2022.77.

Krementsov, Nicolai (1997): *Stalinist Science*, Princeton: Princeton University Press.

Kuhlmann, Wolfgang (1985): *Reflexive Letztbegründung: Untersuchungen zur Transzendentalpragmatik*, Freiburg i.Br.: Alber Verlag.

Kuhlmann, Wolfgang (2010): *Unhintergehbarkeit: Studien zur Transzendentalpragmatik*, Würzburg: Verlag Königshausen & Neumann.

Kuhn, Thomas (1962/1970): *The Structure of Scientific Revolutions*, 2nd ed., Chicago: The University of Chicago Press.

Kuhn, Thomas (1977): "Objectivity, Value Judgment and Theory Choice", in *The Essential Tension*, Chicago: The University of Chicago Press, pp. 320–339.

Lacey, Hugh (1999): *Is Science Value-Free? Values and Scientific Understanding*, New York: Routledge.

Lacey, Hugh (2005): "On the Interplay of the Cognitive and the Social in Scientific Practices", *Philosophy of Science*, vol. 72, pp. 977–988.

Lakatos, Imre (1970): "Falsification and the Methodology of Scientific Research Programmes", in Imre Lakatos and Alan Musgrave (eds.): *Criticism and the Growth of Knowledge*, Cambridge: Cambridge University Press, pp. 91–196.

Landemore, Hélène (2012): "Collective Wisdom: Old and New", in Hélène Landemore and Jon Elster (eds.): *Collective Wisdom: Principles and Mechanisms*, Cambridge: Cambridge University Press, pp. 1–20.

Lange, Marc (2016): *Because without Cause: Non-Causal Explanations in Science and Mathematics*, Oxford: Oxford University Press.

Larkin, Jill H. and Simon, Herbert (1987): "Why a Diagram Is (Sometimes) Worth Ten Thousand Words", *Cognitive Science*, vol. 11, pp. 65–99.

Latour, Bruno, and Woolgar, Steve (1979/1986): *Laboratory Life: The Social Construction of Scientific Facts*, 2nd ed., Princeton: Princeton University Press.

Laudan, Larry (1977): *Progress and Its Problems: Towards a Theory of Scientific Growth*, Los Angeles and London: University of California Press.

Laudan, Larry (1981): "A Problem-Solving Approach to Scientific Progress", in Ian Hacking (ed.): *Scientific Revolutions*, Oxford: Oxford University Press, pp. 144–155.

Laudan, Larry (1984): *Science and Values*, Berkeley; Los Angeles; London: University of California Press.

Laudan, Larry (1987): "Progress or Rationality? The Prospects for Normative Naturalism", *American Philosophical Quarterly*, vol. 24, pp. 19–31.

Laudan, Larry (1990): *Science and Relativism*, Chicago and London: The University of Chicago Press.

Laudan, Larry (2004): "The Epistemic, the Cognitive, and the Social", in Peter Machamer and Gideon Wolters (eds.): *Science, Values and Objectivity*, Pittsburgh: University of Pittsburgh Press, pp. 14–23.

Leonard, Thomas C. (2002): "Reflections on Rules in Science: An Invisible-Hand Perspective", *Journal of Economic Methodology*, vol. 9, pp. 141–168.

Levi, Isaac (1960): "Must the Scientist Make Value Judgments?", *Journal of Philosophy*, vol. LVII, pp. 345–357.

Levi, Isaac (1962): "On the Seriousness of Mistakes", *Philosophy of Science*, vol. 29, pp. 47–65.

Levi, Margaret (1988): *Of Rule and Revenue*, Berkeley; Los Angeles; London: University of California Press.

Lewis, John David (2011): "Constitutional and Fundamental Law: The Lesson of Classical Greece", in Ellen Frankel Paul, Miller Fred D., Jr., and Paul Jeffrey (eds.): *What Should Constitutions Do?*, Cambridge: Cambridge University Press, pp. 25–49.

Lindquist, K., Wager, T. D., Kober, H., Bliss-Moreau, E., and Barrett, L. F. (2012): "The Brain Basis of Emotion: A Meta-Analytic Review", *Behavioral and Brain Sciences*, vol. 35(3), pp. 121–143.

List, Christian and Pettit, Philip (2011): *Group Agency*, Oxford: Oxford University Press.

Locke, John (1690): *An Essay Concerning Human Understanding*, London: Thomas Basset.

Longino, Helen (1990): *Science as Social Knowledge*, Princeton: Princeton University Press.

Longino, Helen (2002): *The Fate of Knowledge*, Princeton: Princeton University Press.

Longino, Helen (2013): *Studying Human Behavior*, Chicago: The University of Chicago Press.

Lorenz, Konrad (1941): "Kants Lehre vom Apriorischen im Lichte gegenwärtiger Biologie", *Blätter für deutsche Philosophie*, vol. 15(1), pp. 95–124.

Loughlin, Martin (2022): *Against Constitutionalism*, Cambridge, MA: Harvard University Press.

Love, Alan C. (2015): "Collaborative Explanation, Explanatory Roles and Scientific Explaining in Practice", *Studies in History and Philosophy of Science*, vol. 52, pp. 88–94.

Machamer, Peter, Darden, Lindley, and Craver, Carl (2000): "Thinking about Mechanisms", *Philosophy of Science*, vol. 67, pp. 1–25.

MacFie, Alec (1971): "The Invisible Hand of Jupiter", *Journal of the History of Ideas*, vol. 32, pp. 595–599.

Magnus, P. D. (2013): "What Scientists Know Is Not a Function of What Scientists Know", *Philosophy of Science*, vol. 80, pp. 840–849.

Magnus, P. D. (2022): "The Scope of Inductive Risk", *Metaphilosophy*, vol. 53, pp. 17–24.

Mahoney, James and Thelen, Kathleen (eds.) (2015): *Advances in Comparative-Historical Analysis*, Cambridge: Cambridge University Press.

Malpighi, Marcello (1661/1929): *De Pulmonibus epistolae II ad Borellium*, 1661, English translation by James Young: Malpighi's "De Pulmonibus", *Proceedings of the Royal Society of Medicine*, vol. 23, 1929, pp. 1–11.

Malzbender, T. and Gelb, D. (2006): *Polynomial Texture Mapping, Hewlett-Packard Mobile and Media Systems Laboratory*. www.hpl.hp.com/research/ptm/.

Mantzavinos, C. (2001): *Individuals, Institutions, and Markets*, Cambridge: Cambridge University Press.

Mantzavinos, C. (2005): *Naturalistic Hermeneutics*, Cambridge: Cambridge University Press.

Mantzavinos, C. (2013): "Explanatory Games", *Journal of Philosophy*, vol. CX, pp. 602–632.

Mantzavinos, C. (2014): "Text Interpretation as a Scientific Activity", *Journal for General Philosophy of Science*, vol. 45, pp. 45–58.

Mantzavinos, C. (2016): *Explanatory Pluralism*, Cambridge: Cambridge University Press.

Mantzavinos, C. (2021a): "Science, Institutions, and Values", *European Journal of Philosophy*, vol. 29, pp. 379–392.

Mantzavinos, C. (2021b): "Institutions and Scientific Progress", *Philosophy of the Social Sciences*, vol. 51, pp. 243–265.

Mantzavinos, C., North, Douglass, and Shariq, Syed (2004): "Learning, Institutions, and Economic Performance", *Perspectives on Politics*, vol. 2, pp. 75–84.

March, James (1999): *The Pursuit of Organizational Intelligence*, Oxford: Blackwell.

McAdams, Richard (1997): "The Origin, Development, and Regulation of Norms", *Michigan Law Review*, vol. 96(1997), pp. 338–433.

McCauley, Robert (2000): "The Naturalness of Religion and the Unnaturalness of Science", in Frank Keil and Robert Wilson (eds.): *Explanation and Cognition*, Cambridge, MA: The MIT Press, pp. 61–85.

McDowell, John (1998): "Virtue and Reason", in *Mind, Value, and Reality*, Cambridge, MA: Harvard University Press, pp. 50–73.

McIlwain, Charles Howard (1940): *Constitutionalism: Ancient and Modern*, Ithaca, NY: Cornell University Press.

McMullin, Ernan (1983/2012): "Values in Science", in Peter D. Asquith and Thomas Nickles (eds.): *Proceedings of the 1982 Biennial Meeting of the Philosophy of Science Association*, 1, East Lansing, MI: Philosophy of Science Association, pp. 3–28, reprinted in: *Zygon*, vol. 47, 2012, pp. 686–709.

Meier, Georg Friedrich (1757/1996): *Versuch einer allgemeinen Auslegungskunst, mit einer Einleitung und Anmerkungen*, (eds.) Axel Bühler and Luigi Cataldi Madonna, Hamburg: Felix Meiner Verlag.

Menger, Carl (1871/1968): *Grundsätze der Volkswirtschaftslehre*, Tübingen: J.C.B. Mohr (Paul Siebeck).

Menger, Carl (1871/1976): *Principles of Economics*, Arlington, VA: The Institute for Humane Studies.

Mercier, Hugo and Sperber, Dan (2017): *The Enigma of Reason*, Cambridge, MA: Harvard University Press.

Merton, Robert (1943/1973): "The Normative Structure of Science", in Robert Merton (ed.): *The Sociology of Science*, Chicago: The University of Chicago Press, pp. 268–278.

Meyer, John (2010): "World Society, Institutional Theories, and the Actor", *Annual Review of Sociology*, vol. 36, pp. 1–20.

Mill, John Stuart (1843/1974): *A System of Logic Ratiocinative and Inductive, The Collected Works of John Stuart Mill*, vol. VIII, Toronto: University of Toronto Press.

Montesquieu, Charles de Secondat Baron de (1748/1989): *The Spirit of the Laws*, trans. A. Cohler, B. Miller, and H. Stone, Cambridge: Cambridge University Press.

Morgan, Mary (2012): *The World in the Model: How Economists Work and Think*, Cambridge: Cambridge University Press.

Moe, Terry (2005): "Power and Political Institutions", *Perspectives on Politics*, vol. 3, pp. 215–233.

Mueller, Dennis (2003): *Public Choice III*, Cambridge: Cambridge University Press.

Nagel, Ernest (1961): *The Structure of Science*, London: Routledge & Kegan Paul.

Nagel, Thomas (1986): *The View from Nowhere*, Oxford: Oxford University Press.

Nagel, Thomas (2012): *Mind and Cosmos*, Oxford: Oxford University Press.

Nau, Heino Heinrich (ed.) (1996): *Der Werturteisstreit. Die Äußerungen zur Werturteilsdiskussion im Ausschuß des Vereins für Socialpolitik (1913)*, Marburg: Metropolis Verlag.

Nee, Victor and Brinton, Mary (eds.) (1998): *The New Institutionalism in Sociology*, New York: Russell Sage.

Nee, Victor and Ingram, Paul (1998): "Embeddedness and Beyond: Institutions, Exchange and Social Structure", in Victor Nee and Mary C. Brinton (eds.): *The New Institutionalism in Sociology*, New York: Russell Sage Foundation, pp. 19–45.

Needham, Joseph (1978): *The Shorter Science and Civilization in China* (prepared under the supervision by Colin A. Ronan), vol. 1, Cambridge: Cambridge University Press.

Nehamas, Alexander (1981): "The Postulated Author: Critical Monism as a Regulative Ideal", *Critical Inquiry*, vol. 8, pp. 133–149.

Nehamas, Alexander (1985): "Convergence and Methodology in Science and Criticism", *New Literary History*, vol. 17 on *Philosophy of Science and Literary Theory*, pp. 81–87.

Nehamas, Alexander (1987): "Writer, Text, Work, Author", in A. J. Cascardi (ed.): *Literature and the Question of Philosophy*, Baltimore: John Hopkins University Press, pp. 267–291.

Nelson, Richard (1959): "The Simple Economics of Basic Scientific Research", *Journal of Political Economy*, vol. 67, pp. 297–306.

Nersessian, Nancy (2008): *Creating Scientific Concepts*, Cambridge, MA: The MIT Press.

Newell, Allen and Simon, Herbert (1972): *Human Problem Solving*, Englewood Cliffs, NJ: Prentice Hall.

Nickelsen, Kärin (2022): "Cooperative Division of Cognitive Labour: The Social Epistemology of Photosynthesis Research", *Journal for General Philosophy of Science*, vol. 53, pp. 23–40.

Nickles, Thomas (2021): "Historicist Theories of Scientific Rationality", *The Stanford Encyclopedia of Philosophy* (Spring 2021 Edition), https://plato.stanford.edu/archives/spr2021/entries/rationality-historicist/.

Niiniluoto, Ilka (2014): "Scientific Progress as Increasing Verisimilitude", *Studies in History and Philosophy of Science Part A*, vol. 46, pp. 73–77.

Niiniluoto, Ilka (2023): "The Semantic Approach: Scientific Progress as Increased Truthlikeness", in Yafeng Shan (ed.): *New Philosophical Perspectives on Scientific Progress*, New York and London: Routledge, pp. 27–45.

Nisbett, Richard and Ross, Lee (1980): *Human Inference: Strategies and Shortcomings of Social Judgment*, Englewood Cliffs, NJ: Prentice Hall.

North, Douglass C. (1981): *Structure and Change in Economic History*, New York: W. W. Norton.

North, Douglass C. (1990): *Institutions, Institutional Change and Economic Performance*, Cambridge: Cambridge University Press.

North, Douglass C. (1994): "Economic Performance through Time", *American Economic Review*, vol. 84, pp. 359–368.

North, Douglass C. (2005): *Understanding the Process of Economic Change*, Princeton: Princeton University Press.

North, Douglass C., Wallis, John, and Weingast, Barry (2009): *Violence and Social Orders: A Conceptual Framework for Interpreting Recorded Human History*, Cambridge: Cambridge University Press.

Norton, John (1993): "The Determination of Theory by Evidence: The Case for Quantum Discontinuity, 1900–1915", *Synthese*, vol. 97, pp. 1–31.

Norton, John (2008): "Must Evidence Underdetermine Theory?", in Carrier Martin, Don Howard, and Janet Kourany (eds.): *The Challenge of the Social and the Pressure of Practice: Science and Values Revisited*, Pittsburgh: Pittsburgh University Press, pp. 17–44.

Ober, Josiah (2015): *The Rise and Fall of Classical Greece*, Princeton: Princeton University Press.

Oddie, Graham (2014): "Truthlikeness", in Edward N. Zalta (ed.): *The Stanford Encyclopedia of Philosophy* (Summer 2014 Edition), https://plato.stanford .edu/archives/sum2014/entries/truthlikeness/.

Opp, Karl-Dieter (1982): "The Evolutionary Emergence of Norms", *British Journal of Social Psychology*, vol. 21, pp. 139–149.

Oppenheim, Heinrich Bernhard (1871): "Manchesterschule und Kathedersozialismus", *National-Zeitung*, vol. 7. Dezember, Nr. 573, 24, p. 1.

Ordeshook, Peter (1992): "Constitutional Stability", *Constitutional Political Economy*, vol. 3, pp. 137–175.

Ostrom, Elinor (1990): *Governing the Commons: The Evolution of Institutions for Collective Action*, Cambridge: Cambridge University Press.

Ostrom, Elinor (2005): *Understanding Institutional Diversity*, Princeton: Princeton University Press.

Papineau, David (1999): "Normativitiy and Judgment", *Aristotelian Society Supplementary Volume*, vol. 73, pp. 17–43.

Palmieri, Paolo (2012): "Signals, Cochlear Mechanics and Pragmatism: A New Vista of Human Hearing?", *Journal of Experimental & Theoretical Artificial Intelligence*, vol. 24, pp. 527–548.

Pamuk, Zeynep (2021): *Politics and Expertise. How to Use Science in a Democratic Society*, Princeton: Princeton University Press.

Parfit, Derek (2011): *On What Matters*, Vol. 2, Oxford: Oxford University Press.

Parker, Wendy and Winsberg, Eric (2018): "Values and Evidence: How Models Make a Difference", *European Journal for Philosophy of Science*, vol. 8(125), pp. 125–142.

Parker, Wendey (2020): "Model Evaluation: An Adequacy-for-Purpose Approach", *Philosophy of Science*, vol. 87, pp. 457–477.

Pascal, Blaise (1670/2000): *Pensées*, Paris: Librairie Générale Française.

Penn, Derek C. and Povinelli, Daniel J. (2007): "On the Lack of Evidence that Non-human Animals Possess Anything Remotely Resembling a 'Theory of Mind'", *Philosophical Transactions of the Royal Society, B*, vol. 362, pp. 731–744.

Perini, Laura (2005a): "The Truth in Pictures", *Philosophy of Science*, vol. 72, pp. 262–285.

Perini, Laura (2005b): "Visual Representations and Confirmation", *Philosophy of Science*, vol. 72, pp. 913–926.

Pettit, Philip (1997): *Republicanism: A Theory of Freedom and Government*, Oxford: Oxford University Press.

Pettit, Philip (2012): *On the People's Terms: A Republican Theory and Model of Democracy*, Cambridge: Cambridge University Press.

Picavet, Emmanuel (2020): "Ways of Compromise-Building in a World of Institutions", in Andina, Tiziana and Peter Bojanic (eds.): *Institutions in Action. The Nature and Role of Institutions in the Real World*, Heidelberg and New York: Springer, pp. 135–145.

Pierson, Paul (2004): *Politics in Time: History, Institutions and Social Analysis*, Princeton: Princeton University Press.

Plutynski, Anya (2018): *Explaining Cancer: Finding Order in Disorder*, Oxford: Oxford University Press.

Poincaré, Henri (1919): *La Morale et la Science: Dernières Pensées*, Paris: Ernest Flammarion.

Polanyi, Michael (1958): *Personal Knowledge*, London: Routledge.

Polanyi, Michael (1962): "The Republic of Science", *Minerva*, vol. I, pp. 54–73.

Polanyi, Michael (1966): *The Tacit Dimension*, Gloucester, MA: Peter Smith.

Pollock, Ethan (2008): *Stalin and the Soviet Science Wars*, Princeton: Princeton University Press.

Popper, Karl (1934/2003): *The Logic of Scientific Discovery*, London and New York: Routledge.

Popper, Karl (1944/1957): *The Poverty of Historicism*, first published in *Economica*, N.S., vol. XI, no. 42 and 43, 1944 and vol. XII, no 46, 1945 and appeared as a book in London: Routledge.

Popper, Karl (1945/2002): *The Open Society and Its Enemies*, London and New York: Routledge.

Popper, Karl (1963): *Conjectures and Refutations: The Growth of Scientific Knowledge*, London and New York: Routledge.

Popper, Karl (1972): *Objective Knowledge: An Evolutionary Approach*, Oxford: Oxford University Press.

Popper, Karl (1994): *The Myth of the Framework: In Defence of Science and Rationality*, London and New York: Routledge.

Potochnik, Angela (2017): *Idealization and the Aims of Science*, Chicago: The University of Chicago Press.

Price, Derek de Solla Price (1974): "Gears from the Greeks: The Antikythera Mechanism – A Calendar Computer from ca. 80 B.C.", *Transactions of the American Philosophical Society New Series*, vol. 64, pp. 1–70.

Pryke, Louise (2016): "Religion and Humanity in Mesopotamian Myth and Epic", *Oxford Research Encyclopedia of Religion*, https://doi.org/10.1093/acref ore/9780199340378.013.247.

Psillos, Stathis (1999): *Scientific Realism: How Science Tracks Truth*, London: Routledge.

Psillos, Stathis (2015): "Evidence: Wanted, Alive or Dead", *Canadian Journal of Philosophy*, vol. 45, pp. 357–381.

Putnam, Hilary (1981): *Reason, Truth and History*, Cambridge: Cambridge University Press.

Putnam, Hilary (2002): *The Collapse of the Fact/Value Dichotomy and Other Essays*, Cambridge, MA: Harvard University Press.

Quine, Willard van Orman (1951/1980): "Two Dogmas of Empiricism", in *From a Logical Point of View*, 2nd ed., Cambridge, MA: Harvard University Press, pp. 20–46.

Quine, Willard van Orman (1960): *Word and Object*, Cambridge, MA: The MIT Press.

Raatikainen, Panu (2022): "Truth and Theories of Truth", in Piotr Stalmaszczyk, (ed.): *The Cambridge Handbook of the Philosophy of Language*, Cambridge: Cambridge University Press, pp. 217–232.

Radnitzky, Gerhard (1981): "Wertfreiheitsthese: Wissenschaft, Ethik und Politik", in Gerhard Radnitzky and Gunnar Andersson (eds.): *Voraussetzungen und Grenzen der Wissenschaft*, Tübingen: J.C.B. Mohr (Paul Siebeck), pp. 47–126.

Railton, Peter (1981): "Probability, Explanation and Information", *Synthese*, vol. 48, pp. 233–256.

Railton, Peter (1986): "Moral Realism", *Philosophical Review*, vol. 95, pp. 163–227.

Rawls, John (1971): *A Theory of Justice*, Cambridge, MA: Harvard University Press.

Reichenbach, Hans (1938): *Experience and Prediction*, Chicago: The University of Chicago Press.

Reichenbach, Hans (1951): *The Rise of Scientific Philosophy*, Berkeley; Los Angeles: University of California Press.

Reiss, Julian (2015): "A Pragmatist Theory of Evidence", *Philosophy of Science*, vol. 82, pp. 341–362.

Reiss, Julian and Sprenger, Jan (2020): "Scientific Objectivity", in Edward N. Zalta (ed.): *Stanford Encyclopedia of Philosophy*.

Rescher, Nicholas (1997): *Objectivity: The Obligations of Impersonal Reason*, Notre Dame, IN and London: University of Notre Dame Press.

Resnik, David (1998): *The Ethics of Science: An Introduction*, London and New York: Routledge.

Resnik, David and Elliott, Kevin (2019): "Value-Entanglement and the Integrity of Scientific Research", *Studies in History and Philosophy of Science Part A*, 75, pp. 1–11.

Resnik, David and Elliott, Kevin (2023): "Science, Values, and the New Demarcation Problem", *Journal for General Philosophy of Science*, published online 22 February 2023.

Rice, Collin (2021): *Leveraging Distortions: Explanation, Idealization and Universality in Science*, Cambridge, MA: The MIT Press.

Rickert, Heinrich (1929): *Die Grenzen der naturwissenschaftlichen Begriffsbildung: Eine logische Einleitung in die historischen Wissenschaften*, 5. Auflage, Tübingen: J.C.B. Mohr (Paul Siebeck).

Riker, William (1980): "Implications from the Disequilibrium of Majority Rule for the Study of Institutions", *American Political Science Review*, vol. 74, pp. 432–447.

Rolin, Kristina (2015): "Values in Science: The Case of Scientific Collaboration", *Philosophy of Science*, vol. 82, pp. 157–177.

Rorty, Richard (1979/2018): *Philosophy and the Mirror of Nature*, Princeton: Princeton University Press.

Rosenfeld, Tobias (2004): "Is Knowing-How Simply a Case of Knowing-That?", *Philosophical Investigations*, vol. 27, pp. 370–379.

Rothschild, Emma (1994): "Adam Smith and the Invisible Hand", *American Economic Review* (Papers and Proceedings), vol. 84, pp. 319–322.

Rothschild, Emma (2002): *Economic Sentiments*, Cambridge, MA: Harvard University Press.

Rowbottom, Darrell (2014): "Aimless Science", *Synthese*, vol. 191, pp. 1211–1221.

Rudner, Richard (1953): "The Scientist qua Scientist Makes Value Judgments", *Philosophy of Science*, vol. 20, pp. 1–6.

Ryle, Gilbert (1949): *The Concept of Mind*, London: Penguin.

Rysiew, Patrick (2021): "Epistemic Contextualism", in Edward N. Zalta (ed.): *Stanford Encyclopedia of Philosophy*.

Salmon, Wesley (1984): *Scientific Explanation and the Causal Structure of the World*, Princeton, NJ: Princeton University Press.

Sankey, Howard (2019): "Objectivity in Science", in: *PhilPapers.*

Sankey, Howard (2021): "Realism and the Epistemic Objectivity of Science", *Kriterion – Journal of Philosophy*, vol. 35, pp. 5–20.

Schiffer, Stephen (2002): "Amazing Knowledge", *Journal of Philosophy*, vol. 99, pp. 200–202.

Schindler, Samuel (2018): *Theoretical Virtues in Science: Uncovering Reality through Theory*, Cambridge: Cambridge University Press.

Schopenhauer, Arthur (1851): "Über Religion: Ein Dialog", in *Parerga and Paralipomena*, §174, Kapitel 15, Berlin: Verlag A. W. Hayn.

Schrödinger, Erwin (1958): *Mind and Matter*, Cambridge: Cambridge University Press.

Schroeder, Andrew S. (2017): "Using Democratic Values in Science: An Objection and (Partial) Response", *Philosophy of Science*, vol. 84, pp. 1044–1054.

Schroeder, Andrew S. (2021): "Democratic Values: A Better Foundation for Public Trust in Science", *British Journal for Philosophy of Science*, vol. 72, pp. 545–562.

Schroeder, Andrew S. (2022): "The Limits of Democratizing Science: When Scientists Should Ignore the Public", *Philosophy of Science*, forthcoming.

Schroeder, Mark (2008): *Being For: Evaluating the Semantic Program of Expressivism*, Oxford: Oxford University Press.

Schurz, Gerhard and Carrier, Martin (eds) (2013): *Werte in den Wissenschaften. Neue Ansätze zum Werturteilsstreit*, Frankfurt am Main: Suhrkamp.

Scriven, Michael (1972): "The Exact Role of Value Judgments in Science", *PSA: Proceedings of the Biennial Meeting of the Philosophy of Science Association*, pp. 219–247.

Sen, Amartya (2006): "What Do We Want from a Theory of Justice?", *Journal of Philosophy*, vol. CIII, pp. 175–208.

Sen, Amartya (2009): *The Idea of Justice*, Cambridge, MA: Harvard University Press.

Servetus, Miguel (1553/2008): *Christianismi Restitutio*, Vienne, English translation: *The Restoration of Christianity*, by Christopher A. Hoffmann and Marian Hiller, Lewiston, NY: The Edwin Mellen Press.

Shackle, G. L. S. (1979): *Imagination and the Nature of Choice*, Edinburgh: Edinburgh University Press.

Shan, Yafeng (2019): "A New Functional Approach to Scientific Progress", *Philosophy of Science*, vol. 86, pp. 739–758.

Shan, Yafeng (2023): "The Functional Approach. Scientific Progress as Increased Usefulness", in Yafeng Shan (ed.): *New Philosophical Perspectives on Scientific Progress*, New York and London: Routledge, pp. 46–61.

Shapin, Steven (1996): *The Scientific Revolution*, Chicago: The University of Chicago Press.

Shepsle, Kenneth (1986): "Institutional Equilibrium and Equilibrium Institutions", in Herbert Weisberg (ed.): *Political Science: The Science of Politics*, New York: Agathon, pp. 51–81.

Shepsle, Kenneth (1989): "Studying Institutions: Some Lessons from the Rational Choice Approach", *Journal of Theoretical Politics*, vol. 1, pp. 131–147.

Shepsle, Kenneth (2006): "Rational Choice Institutionalism", in R. A. W. Rhodes, Sarah Binder, and Bert Rockman (eds.): *Oxford Handbook of Political Institutions*, Oxford: Oxford University Press, pp. 23–38.

Simon, Herbert (1983): *Reason in Human Affairs*, Stanford, CA: Stanford University Press.

Siraisi, Nancy (1981): *Tadeo Alderotti and His Pupils: Two Generations of Italian Medical Learning*, Princeton: Princeton University Press.

Skinner, Quentin (1969): "Meaning and Understanding in the History of Ideas", *History and Theory*, vol. 8, pp. 3–53.

Skinner, Quentin (1972): "Motives, Intentions and the Interpretation of Texts", *New Literary History*, vol. 3, pp. 393–408.

Skinner, Quentin (1975): "Hermeneutics and the Role of History", *New Literary History*, vol. 7, pp. 209–232.

Skow, Bradford (2018): *Causation, Explanation, and the Metaphysics of Aspect*, Oxford: Oxford University Press.

Smith, Adam (1759/1976): *The Theory of Moral Sentiments*, Oxford: Oxford University Press.

Smith, Adam (1776/1976): *An Inquiry into the Nature and Causes of the Wealth of Nations*, (ed.) Edwin Cannan, Chicago: The University of Chicago Press.

Smith, Adam (1759/1980): *Essays in Philosophical Subjects*, Oxford: Oxford University Press.

Smith, Michael (1994): *The Moral Problem*, Oxford: Blackwell.

Snell, Bruno (1946/2009): *Die Entdeckung des Geistes: Studien zur Entstehung des europäischen Denkens bei den Griechen*, 9th ed., Göttingen: Vandnhoeck & Ruprecht.

Snow, P. C. (1959/1993): *The Two Cultures*, Cambridge: Cambridge University Press.

Sober, Elliott (2002): "What Is the Problem of Simplicity?" in Arnold Zellner, Hugo Keuzenbach, and Michael McAleer (eds.): *Simplicity, Inference and Econometric Modelling*, Cambridge: Cambridge University Press, pp. 13–31.

Sokal, Alan (1996): "Transgressing the Boundaries: Towards a Transformative Hermeneutics of Quantum Gravity", *Social Text*, vol. 46/47, pp. 217–252.

Sokal, Alan and Bricmont, Jean (1997): *Inpostures Intellectuelles*, Paris: Éditions Odile Jacob.

Stanley, Jason (2011): *Know How*, Oxford: Oxford University Press.

Stanley, Jason and Williamson, Timothy (2001): "Knowing How", *Journal of Philosophy*, vol. 98, pp. 411–444.

Steel, Daniel (2010): "Epistemic Values and the Argument from Inductive Risk", *Philosophy of Science*, vol. 77, pp. 14–34.

Steele, Katie (2012): "The Scientist qua Policy Advisor Makes Value Judgments", *Philosophy of Science*, vol. 79, pp. 893–904.

Sterelny, Kim (2012): *The Evolved Apprentice: How Evolution Made Humans Unique*, Cambridge, MA: The MIT Press.

Stevenson, Charles Leslie (1937): "The Emotive Meaning of Ethical Terms", *Mind*, vol. 46, pp. 14–31.

Street, Sharon (2006): "A Darwinian Dilemma for Realist Theories of Value", *Philosophical Studies*, vol. 127, pp. 109–166.

Strevens, Michael (2006a): "The Role of the Priority Rule in Science", *Journal of Philosophy*, vol. C, pp. 55–79.

Strevens, Michael (2006b): "The Role of the Matthew Effect in Science", *Studies in History and Philosophy of Science*, vol. 37, pp. 159–170.

Strevens, Michael (2008): *Depth: An Account of Scientific Explanation*, Cambridge, MA: Harvard University Press.

Strevens, Michael (2020): *The Knowledge Machine: How an Unreasonable Idea Created Modern Science*, New York: Penguin Books.

Stump, David (2007): "Pierre Duhem's Virtue Epistemology", *Studies in History and Philosophy of Science*, vol. 38, pp. 149–159.

Stump, David (2011): "The Scientist as Impartial Judge: Moral Values in Duhem's Philosophy of Science: New Perspectives on Pierre Duhem's *The Aim and Structure of Physical Theory*" (book symposium), in: *Metascience*, vol. 20, pp. 1–25.

Sturgeon, Nicholas (1988): "Moral Explanations", in Geoffrey Sayre-McCord (ed.): *Essays on Moral Realism*, Ithaca: Cornell University Press, pp. 229–255.

Svoronos, Ioannes N. (1908): *Das Athener Nationalmuseum*, Athens: Beck & Barth.

Teller, Paul (2008): "Representation in Science", in Stathis Psillos and Martin Curd (eds.): *The Routledge Companion to Philosophy of Science*, London and New York: Routledge, pp. 435–441.

Thagard, Paul (2012): *The Cognitive Science of Science: Explanation, Discovery and Conceptual Change*, Cambridge, MA: The MIT Press.

Thelen, Kathleen (2004): *How Institutions Evolve: The Political Economy of Skills in Germany, Britain, the United States and Japan*, Cambridge: Cambridge University Press.

Tomasello, Michael (2016): *A Natural History of Human Morality*, Cambridge, MA: Harvard University Press.

Topitsch, Ernst (1958): *Vom Ursprung und Ende der Metaphysik: Eine Studie zur Weltanschauungskritik*, Wien: Springer Verlag.

Trout, J. D. (2016): *Wondrous Truths: The Improbable Triumph of Modern Science*, Oxford: Oxford University Press.

Tsebelis, George (2002): *Veto Players: How Political Institutions Work*, Princeton: Princeton University Press.

Tsebelis, George (2017): "The Time Inconsistency of Long Constitutions", *Summer Bulletin of American Academy of Arts and Sciences*, pp. 42–45.

Tullock, Gordon (1987): *Autocracy*, Dordrecht: Kluwer.

Ullmann-Margalit, Edna (1977): *The Emergence of Norms*, Oxford: Clarendon Press.

Ullmann-Margalit, Edna (1978): "Invisible-Hand Explanations", *Synthese*, vol. 39, pp. 263–291.

Van Fraassen, Bas (1980): *The Scientific Image*, Oxford: Oxford University Press.

Van Fraassen, Bas (2008): *Scientific Representation*, Oxford: Oxford University Press.

Vesalius, Andreas (1543): *De Humani Corporis Fabrica Libri Septem*, Basel: Johannes Oporinus, English translation: *On the Fabric of the Human Body*, by William Frank Richardson and John Burd Carman, *vol. 5 containing Books VI: The Heart and Associated Organs and Book VI: The Brain* (Novato CA: Norman Anatomy Series, No 5).

Virvidakis, Stelios (1996): *La robustesse du bien: Essai sur le réalisme moral*, Paris: Editions Jacqueline Chambon.

Voigt, Stefan (1999): *Explaining Constitutional Change*, Aldershot: Edward Elgar, 1999.

Voigt, Stefan (2020): *Constitutional Economics*, Cambridge: Cambridge University Press.

Wallis, Charles (2009): "Consciousness, Context, and Know-How", *Synthese*, vol. 160, pp. 123–153.

Weber, Max (1904/1985): "Die 'Objektivität' sozialwissenschaftlicher und sozialpolitischer Erkenntnis", *Archiv für Sozialwissenschaft und Sozialpolitik*, 1904 and reprinted in: *Gesammelte Aufsätze zur Wissenschaftslehre*, ed. Johannes Winckelmann, 6th ed., Tübingen: Mohr Siebeck, pp. 146–214.

Weber, Max (1913/1996): "Gutachten zur Werturteilsdiskussion im Ausschuß des Vereins für Sozialpolitik", in Heino Heinrich Nau (ed.): *Der Werturteilsstreit. Die Äußerungen zur Werturteilsdiskussion im Ausschuß des Vereins für Sozialpolitik (1913)*, Marburg: Metropolis Verlag, pp. 147–186.

Weber, Max (1917/1985): "Der Sinn der >>Wertfreiheit<< der soziologischen und ökonomischen Wissenschaften" in Johannes Winckelmann (ed.): *Gesammelte Aufsätze zur Wissenschaftslehre*, 6th ed., Tübingen: Mohr Siebeck, pp. 489–540.

Weber, Max (1917/2004): *Science as a Vocation*, Indianapolis: Hackett Publishing House.

Weber, Max (1922/1992): *Wirschaft und Gesellschaft*, 5th ed., Tübingen: J.C.B. Mohr (Paul Siebeck).

Wedgwood, Ralph (2007): *The Nature of Normativity*, Oxford: Clarendon Press.

Weingast, Barry (1997): "The Political Foundations of Democracy and the Rule of Law", *American Political Science Review*, vol. 91, pp. 245–263.

Weisberg, Michael (2007): "Three Kinds of Idealization", *The Journal of Philosophy*, vol. 104, pp. 639–659.

Weisberg, Michael (2013): *Simulation and Similarity: Using Models to Understand the World*, Oxford: Oxford University Press.

Whitehead, Alfred North (1911): *An Introduction to Mathematics*, London: Williams and Norgate.

Wilholt, Torsten (2009): "Bias and Values in Scientific Research", *Studies in History and Philosophy of Science*, vol. 40, pp. 92–101.

Wilholt, Torsten (2010): "Scientific Freedom: Its Grounds and Their Limitations", *Studies in History and Philosophy of Science*, vol. 41, pp. 174–181.

Wilholt, Torsten (2013): "Epistemic Trust in Science", *British Journal for Philosophy of Science*, vol. 64, pp. 233–253.

Wilholt, Torsten (2014): "Review of Philip Kitcher: Science in a Democratic Society", *Philosophy of Science*, vol. 81, pp. 165–171.

Wilholt, Torsten (2016): "Collaborative Research, Scientific Communities, and the Social Diffusion of Trustworthiness", in Michael Brady and Miranda Fricker (eds.): *The Epistemic Life of Groups: Essays in the Epistemology of Collectives*, Oxford: Oxford University Press, pp. 218–233.

Wilholt, Torsten (2022): "Epistemic Interests and the Objectivity of Inquiry", *Studies in History and Philosophy of Science*, vol. 91, pp. 86–93.

Williams, Bernard (1966/1973): "Consistency and Realism", *Proceedings of the Aristotelian Society*, Suppl. Volume 49 (1966), reprinted in *Problems of the Self*, Cambridge: Cambridge University Press, 1973, pp. 187–206.

Williams, Bernard (1985): *Ethics and the Limits of Philosophy*, Cambridge, MA: Harvard University Press.

Williamson, Oliver (1985): *The Economic Institutions of Capitalism*, New York: Free Press.

Williamson, Oliver (1996): *The Mechanisms of Governance*, Oxford: Oxford University Press.

Wimsatt, William (2007): *Re-Engineering Philosophy for Limited Beings: Piecewise Approximations to Reality*, Cambridge, MA: Harvard University Press.

Windelband, Wilhelm (1894/1915): "Geschichte und Naturwissenschaft", in Wilhelm Windelband (ed.): *Präludien: Aufsätze und Reden zur Philosophie und ihrer Geschichte*, Band 2, 5. erweiterte Auflage, Tübingen: J.C.M. Mohr (Paul Siebeck), pp. 136–160.

Woodward, James (2003): *Making Things Happen*, Oxford: Oxford University Press.

Wittgenstein, Ludwig (1922): *Tractatus Logico-Philosophicus*, Translated from the German by C. K. Ogden, London: Routledge & Kegan Paul.

Wodak, Daniel (2017): "Expressivism and Varieties of Normativity", *Oxford Studies in Metaethics*, vol. 12, pp. 265–293.

Woody, Andrea (2015): "Re-orienting Discussions of Scientific Explanation: A Functional Perspective", *Studies in History and Philosophy of Science*, vol. 52, pp. 79–87.

Wootton, David (2015): *The Invention of Science: A New History of Scientific Revolution*, London: Penguin.

Wright, Cory (2015): "The Ontic Conception of Scientific Explanation", *Studies in History and Philosophy of Science*, vol. 54, pp. 20–30.

Zamora-Bonilla, Jesus (2002): "Scientific Inference and the Pursuit of Fame: A Contractarian Approach", *Philosophy of Science*, vol. 69, pp. 300–32.

Zamora-Bonilla, Jesus (2013): "Cooperation, Competition, and the Contractarian View", *Ethics and Politics*, vol. 15, pp. 14–24.

Zollman, Kevin (2018): "The Credit Economy and the Economic Rationality of Science", *Journal of Philosophy*, vol. CXV, pp. 5–33.

Index

Milton Keynes UK
Ingram Content Group UK Ltd.
UKHW031512081224
452171UK00012B/83